Radar Hydrology

PRINCIPLES, MODELS, AND APPLICATIONS

Radar Hydrology

PRINCIPLES, MODELS, AND APPLICATIONS

YANG HONG • JONATHAN J. GOURLEY

CRC Press
Taylor & Francis Group
Boca Raton London New York

CRC Press is an imprint of the
Taylor & Francis Group, an **informa** business

CRC Press
Taylor & Francis Group
6000 Broken Sound Parkway NW, Suite 300
Boca Raton, FL 33487-2742

First issued in paperback 2018

© 2015 by Taylor & Francis Group, LLC
CRC Press is an imprint of Taylor & Francis Group, an Informa business

No claim to original U.S. Government works

ISBN 13: 978-1-138-85536-6 (pbk)
ISBN 13: 978-1-4665-1461-4 (hbk)

Visit the Taylor & Francis Web site at
http://www.taylorandfrancis.com

and the CRC Press Web site at
http://www.crcpress.com

Contents

Preface

The origins of radar date back to World War II, when they provided a new and unique capability to detect enemy aircraft, submarines on the ocean surface, and ships. Radar not only changed the face of the war, but once it became adapted to observe the weather, it led to a revolution in meteorology. It has been instrumental in the study of severe thunderstorms, identification of rotation associated with mesocyclones and tornadoes, detection of severe hail and damaging winds, and estimation of heavy rainfall associated with flash floods. For these reasons, many countries throughout the world have invested in large radar networks for routine observations used to warn the public of these imminent weather hazards.

This book focuses on the use of radars in hydrology. Weather radars have proven their value for remote sensing of precipitation, even at high enough resolution to monitor and predict the onset of flash floods. But the process to arrive at an accurate estimate of precipitation from the raw radar signal is not a straightforward one. For this reason, five chapters of this book are dedicated to radar-based precipitation estimation alone. Graduate students, operational forecasters, and researchers will acquire the theoretical framework and practical experience behind radar precipitation estimation.

We present new radar technologies that will improve the accuracy and resolution of precipitation estimates. The description of these platforms, some of which are mobile or transportable, does not attempt to comprehensively cover all new radar technologies. Rather, we focus on platforms that are more familiar to the authors. Likewise, several of the studies we highlight reflect our own experiences with those observing platforms, basins, and methodologies. We supply complete bibliographies and encourage the interested reader to explore those other studies to gain a more holistic understanding of the topics presented herein.

We believe the next revolution in hydrology will be initiated by radar remote sensing of additional variables going beyond precipitation. Space-based, airborne, and ground-based radars operating at multiple frequencies can be used to detect and measure surface water spatial extent and depth, stream discharge, near-surface soil moisture, subsurface water, and depth to the water table. Radars are now providing insights into water storage and fluxes in regions that have only scarcely been observed. These new observations will influence new hydrologic theories, formulations, and basic understanding. Moreover, accurate estimation of the freshwater storage on Earth will provide the pulse of the planet's climate state.

Acknowledgments

We wish to express our sincere appreciation for the contributions of many individuals who helped write, copyedit, proofread, and review several components of this book. The idea of writing a book originated from a graduate class "Radar Hydrology" taught by the co-authors at the University of Oklahoma. The completion of this book would not have been possible without all the fruitful discussions and contributions from members of the Hydrometeorology and Remote Sensing Laboratory at the University of Oklahoma (http://hydro.ou.edu). Particularly, we are indebted to the following graduate students and research scientists who helped in the preparation of specific chapters and figures, collection of references, and development of problem sets: Dr. Qing Cao, Race Clark, Zac Flamig, Dr. Pierre-Emmanuel Kirstetter, Humberto Vergara, Yixin "Berry" Wen, Amanda Oehlert, Zhanming Wan, Xiaodi Yu, and Yu Zhang. Their assistance and expertise are greatly appreciated and acknowledged!

I (Dr. Hong) would like to thank Drs. Soroosh Sorooshian, Kuolin-Hsu and Robert F. Adler for selflessly sharing their knowledge and wisdom in science and engineering. I also wanted to acknowledge my colleagues and students at the University of Oklahoma and also the National Weather Center for helping me to find my inner passion as well as my niche in the exciting but fast-paced academic world. I dedicate this book to my parents, my family and friends.

I (Dr. Gourley) wanted to first acknowledge Robert A. Maddox for sharing his passion in science and teaching me to strive to become a meticulous researcher. Kenneth W. Howard gave me an opportunity as a fledgling undergraduate student at the National Severe Storms Laboratory I didn't think I deserved. I admire his approach that blends art and science to continually create and invent new ideas. I dedicate this book to my loving family, Steph, Joe, and Gigi. They are my raison d'être!

Finally, we wish to give credit to the numerous funding agencies, including NOAA, NASA, NSF and the University Strategic Organization Advanced Radar Research Center at the University of Oklahoma, for promoting scientific research and engineering development in radar hydrology. We also acknowledge the management and editorial assistance of Ashley Gasque and Andrea Dale.

Drs. Yang Hong and Jonathan J. Gourley
Director and Co-Director of HyDROS Lab
National Weather Center, Norman, Oklahoma

About the Authors

Dr. Yang Hong is a professor of hydrometeorology and remote sensing in the School of Civil Engineering and Environmental Sciences and adjunct faculty with the School of Meteorology, University of Oklahoma. Previously, he was a research scientist at NASA's Goddard Space Flight Center and postdoctoral researcher at University of California, Irvine.

Dr. Hong currently directs the HyDROS Lab (Hydrometeorology and Remote Sensing Laboratory: http://hydro.ou.edu) at the National Weather Center and also serves as the co-director of WaTER (Water Technology for Emerging Regions) Center, faculty member with the Advanced Radar Research Center, and affiliated member of the Center for Analysis and Prediction of Storms at the University of Oklahoma. Dr. Hong's areas of research span the wide range of hydrology–meteorology–climatology, with particular interest in bridging the gap among the water–weather–climate–human systems across scales in space and time. He has developed and taught class topics such as remote sensing retrieval and applications, advanced hydrologic modeling, climate change and natural hazards, engineering survey/measurement and statistics, land surface modeling and data assimilation systems for hydrological cycle and water systems under a changing climate.

Dr. Hong has served on several international and national committees, review panels, and editorial boards of several journals. He has served as chair of the AGU-Hydrology Section Technique Committee on Precipitation (2008–2012) and as editor for numerous journals. He received the 2008 Group Achievement Award from NASA "for significant achievements in systematically promoting and accelerating the use of NASA scientific research results for societal benefits," and the 2014 University of Oklahoma Regents' Award for Superior Research "for superior accomplishments in teaching, research and creative activity, and professional and university service." He has extensively published in journals of remote sensing, hydrology, meteorology, and hazards and has released several technologies to universities, governmental agencies, and private companies.

Dr. Hong earned a PhD in hydrology and water resources with a minor in remote sensing and spatial analysis from the University of Arizona (2003) and an MS (1999) in environmental sciences and a BS (1996) in geosciences from Peking (Beijing) University, China.

Dr. Jonathan Gourley is a research hydrologist with the NOAA/National Severe Storms Laboratory and an affiliate associate professor with the School of Meteorology at the University of Oklahoma in Norman. He earned his BS and MS degrees in meteorology and PhD in civil engineering from the University of Oklahoma. His primary research interests include hydrologic prediction across scales ranging from water resources management down to early warning of extreme events such as flash floods, rainfall estimation from remote-sensing platforms such as satellite-based measurements and dual-polarization radar, and improving theoretical understanding of water fluxes through unique observations. He was instrumental in proving the usefulness of dual-polarization radar during his postdoctoral research in France, and MètèoFrance has subsequently upgraded their network with this capability.

Dr. Gourley was the principal inventor of a multisensor rainfall algorithm that has now been expanded to encompass all radars in North America and has been deployed to several foreign countries for operational use. He assembled a comprehensive database of flooding in the United States for community research purposes based on stream gauges, trained spotter reports, and witness reports collected directly from the public through an experiment he coordinated. This database is being used to develop and evaluate a modeling system called FLASH (Flooded Locations and Simulated Hydrographs), which is being designed to provide state-of-the-art forecasts of flash flooding in real time over the United States. Dr. Gourley has won Department of Commerce Bronze and Silver Medal Awards as well as the American Meteorological Society *Journal of Hydrometeorology* Editor's Award. More information about his educational background and research can be found at http://blog.nssl.noaa.gov/jjgourley/ and at http://blog.nssl.noaa.gov/flash/.

1

Introduction to Basic Radar Principles

Radar is an acronym for *Ra*dio *D*etection *a*nd *R*anging. A British invention, it was initially designed to detect aircraft, warships, and surfacing submarines, and its successful implementation ultimately shaped the outcome of the Second World War. Since then, the use of radar technologies has significantly expanded beyond military to civilian and commercial applications. Air traffic controllers use them to direct aircraft and avoid collisions. Police officers routinely use Doppler radars to detect speeding cars down the road. Even the microwave oven used to heat up a cup of coffee is a result of radar! Radar observations of variables in the hydrologic cycle have led to a panacea of new discoveries as well as monitoring and forecasting capabilities that have greatly impacted the field of hydrology. This chapter provides an introduction to basic radar principles. While the theories and equations are universal, the myriad types of radars nowadays have very diverse applications and operating characteristics. Weather radar applications are diverse and far-reaching, but the focus in this chapter is on the hydrologic use of weather radar. In many cases, examples and typical values for the variables will be provided for the Weather Surveillance Radar–1988 Doppler (WSR-88D)—the radar that constitutes the Next Generation Radar (NEXRAD) network in operation across the United States.

1.1 Radar Components

Radar is an instrument that consists of the basic components shown in Figure 1.1. The **transmitter** generates electromagnetic (EM) radiation as a pulse or continuous wave. The WRD-88D radar employs a klystron transmitter to generate a pulse of energy; the klystron transmitter is typically more expensive than a magnetron due to its ability to control the frequency of the transmitted signal. Chapter 3 discusses the advantages of a radar that transmits and receives signals that are polarized in both the horizontal and vertical planes. The polarized, transmitted pulse travels through the **waveguide**, which is typically a hollow conduit made of conductive metal with a rectangular cross-section. Some dual-polarized radars are designed with two separate waveguides corresponding to the horizontal and vertical channels. The waveguide connects the transmitter to the radar **antenna**, most commonly

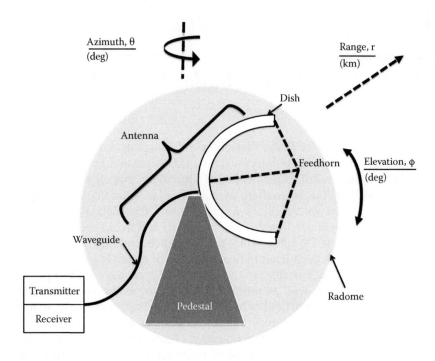

FIGURE 1.1
Basic components of a conventional weather radar system. Spherical coordinates of range (km), azimuth angle (deg), and elevation angle (deg) are most convenient when describing radar data.

consisting of a parabolic **dish**, or **reflector**, that is mechanically rotated by a **pedestal** in the azimuthal direction and vertically in elevation. The pedestal is the primary moving component of a radar system and requires regular maintenance. Chapter 5 introduces a phased-array radar, which does not necessarily require a pedestal. The EM pulse is directed to a device called a **feedhorn**, which conveys the alternating current as a radio wave a short distance to the center of the dish. Here, the radio wave is reflected off the dish to form the radar beam and transmitted through the free atmosphere to the intended target. Some of this energy encounters objects in the atmosphere and is subsequently scattered back to the dish; this energy is called the **backscatter**. The reflector concentrates the backscattered energy to the feedhorn. The feedhorn acts to convert the backscattered radio wave back into a voltage. The received signal travels down the feedhorn to the receiver, where it is amplified and subsequently processed.

The radar antenna, including the parabolic dish, feedhorn, and pedestal, is often protected by a spherical radome. The primary purpose of the radome is to minimize wind loading on the dish, causing excessive strain on the pedestal. The radome also serves to conceal the antenna, protect personnel from moving parts, and to protect the antenna from accumulating snow and ice. The radome is typically constructed from multiple panels that can be either symmetric or asymmetric. The material of the radome needs to be nonconductive and free from metal screws and wires, so as to avoid artifacts in the data and signal loss; fiberglass is most often used. It is also important that the area immediately surrounding the radome remains free from any objects, especially metallic ones. Gourley et al. (2006) examined the data quality of radar variables measured by MeteoFrance's Trappes polarimetric radar. They found artifacts in the radar data that resulted from a security fence mounted along the perimeter of the tower within the field of view of the antenna. Furthermore, they found significant biases in radar variables at specific azimuths and elevation angles that coincided with the location of a small box (20 × 27 × 60 cm³) containing electronics to operate an elevator.

1.2 The Radar Beam

Weather radars transmit EM energy in the microwave spectrum that travel at the speed of light in a vacuum at 3×10^8 ms^{-1}. The relationship between radio frequency (f), wavelength (λ), and velocity at the speed of light (c) is the following:

$$c = f\lambda \tag{1.1}$$

where c is 3×10^8 ms^{-1}, f is in cycles per second, or Hertz (Hz), and λ is in m. Table 1.1 shows the most common bands, frequencies, and associated wavelengths that correspond to radars that have hydrologic applications.

Note that the typical values for radar microwave frequencies are in the order of $10^7 - 10^{11}$ Hz; thus it is convenient to use Mega (10^6) and Giga (10^9) prefixes, or MHz and GHz. The corresponding radar wavelengths span a few millimeters (mm) up to m. The radar wavelength and diameter (d) of the parabolic dish dictate the angular width of the radar beam, or **beamwidth(θ)**, as follows:

$$\theta = \frac{73\lambda}{d} \tag{1.2}$$

where λ and d are both in the same distance units and θ is in deg. In the case of the WSR-88D radar, it operates at an approximate 10.7 cm wavelength and has an 8.5 m diameter dish. This corresponds to a beamwidth of approximately 0.92 deg (in both azimuth and elevation directions). Targets with horizontal cross-sections (for a horizontally polarized wave) less than $\lambda/16$, or approximately 7 mm for the WSR-88D, are Rayleigh scatters and thus have predictable radar signatures for different-sized raindrops. The targets

TABLE 1.1

Summary of Radar Characteristics Used for Hydrologic Applications

Band	Frequency	Wavelength	Hydrologic Applications
W	75–110 GHz	2.7–4.0 mm	Detection of cloud droplets
mm	40–300 GHz	7.5–1 mm	Cloud microphysical processes
Ka	24–40 GHz	0.8–1.1 cm	Precipitation estimation from spaceborne radar, streamflow, surface water heights
Ku	12–18 GHz	1.7–2.5 cm	Precipitation estimation from spaceborne radar, surface water velocity
X	8–12 GHz	2.5–3.8 cm	High-resolution precipitation and microphysical studies, surface water extent and depth
C	4–8 GHz	3.8–7.5 cm	Estimation of light-moderate precipitation, surface water extent and depth, top-layer soil moisture
S	2–4 GHz	7.5–15 cm	Estimation of moderate-heavy precipitation
L	1–2 GHz	15–30 cm	Top-layer soil moisture
UHF	300–1000 MHz	0.3–1 m	Ground-penetrating radar for soil moisture and water table, channel bathymetry
P	~300 MHz	1 m	Root-zone soil moisture
VHF	30–300 MHz	1–10 m	Ground-penetrating radar for soil moisture and water table

are assumed to produce scattering equal in all directions, called **isotropic scattering**. The radar detects the component of scattering that comes back to the radar (backscatter). Shorter wavelength radars at X band and shorter have a lower upper limit on the diameter of targets that cause Rayleigh scattering. But, these smaller wavelength radars do not require such large dishes to maintain a small beamwidth desirable for high-resolution precipitation measurements, and thus are more amenable to spaceborne, transportable, and mobile radar platforms. These shorter radar wavelengths are more prone to absorption of the radar signal by atmospheric gases and precipitation leading to signal loss with increasing range from the radar; this phenomenon is called **attenuation**.

The beamwidth is defined by the angular width at which the power in the center of the beam drops in half; this is the **half-power point** of the beam. Figure 1.2 illustrates how the beam volume (or distance between the upper and lower part of the beam) increases with range; this is referred to as **beam broadening** or **beam spreading**. This characteristic is one of the major limiting factors of radar measurements taken at far range. In addition to beam broadening, the height of the beam's center relative to Earth's surface increases with range. Beam broadening and beam heights increasing with range drastically alter the location, shape, and volume of an individual radar bin.

The beam propagation path depends on atmospheric conditions (primarily the vertical distribution of water vapor pressure and temperature) that control the change in the atmospheric refractive index (N) with height.

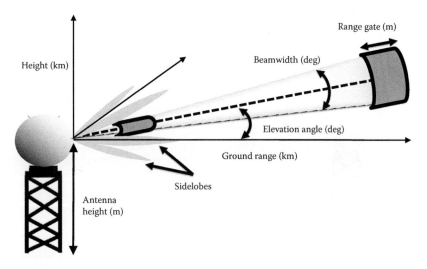

FIGURE 1.2
Variables used to describe the radar beam as it propagates away from the radar through the atmosphere.

Anticipating the precise beam propagation path is difficult and uncertain, as difficult-to-detect changes in the refractive index can occur with commonly occurring atmospheric phenomena such as temperature inversions associated with cold fronts, cool/moist thunderstorm outflows, and nocturnal radiation. These conditions cause the beam to bend or duct downward; this is called **superrefraction** as shown in Figure 1.3. In some cases, the superrefraction is severe enough that the beam ducts down, strikes Earth's surface, and sends a strong signal back to the radar. Echoes from this **anomalous propagation** situation, or **anaprop**, show up on the radar display and can mislead the radar user into believing there are storms nearby when actually there are none. The opposite case of the radar beam ducting away from Earth's surface at a rate that deviates from conditions that occur in a standard atmosphere is called **subrefraction**. This phenomenon is less noticeable to radar users, but can occur in drier regions where the temperature decreases rapidly with height and the relative humidity increases. Atmospheric conditions leading to beam subrefraction are common in the western United States during the warm season. Under standard atmospheric conditions, the relationship between the height (h) of the center of the radar beam above Earth's surface is given as

$$h = h_r + \sqrt{r^2 + \left(\frac{4}{3}a\right)^2 + 2r\frac{4}{3}a\sin\phi_e} - \frac{4}{3}a \qquad (1.3)$$

where h_r is the height of the radar antenna, which can be approximated by the height of the radar tower (km), r is range from the radar (km), a is

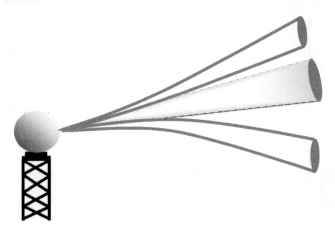

FIGURE 1.3
The radar beam under different atmospheric conditions that result in superrefraction and subrefraction.

Earth's radius or approximately 6371 km, and ϕ_e is the radar elevation angle (deg) (refer to Figure 1.2). Figure 1.2 illustrates how the height of the beam's centerline increases with range from the radar. As shown in more detail in Chapter 2, the increase of the radar beam height (and volume) with range causes it to sample higher in the clouds, causing biases in rainfall estimates.

Approximately 80% of the transmitted power resides within the main lobe of the radar beam, while some of the transmitted power "leaks" outside the main beam into what are referred to as **sidelobes** (see Figure 1.2). The amount of power leaking into the sidelobes depends on the design of the antenna. Some of the EM energy can reflect off the ground, nearby buildings, and trees resulting in **ground clutter** contamination. These artifacts can be associated with significant backscattered energy, leading to misinterpretations of the radar signal and biases in derived precipitation fields. They appear predominantly at ranges near the radar where the main lobe and sidelobes are close to the surface. Numerous algorithms have been developed to identify and screen out radar echoes that are associated with ground clutter. The most effective techniques for discriminating nonweather echoes rely on data from dual-polarization radar, discussed in more detail in Chapter 3.

For a radar to estimate spatially and temporally variable precipitation fields, it must transmit and receive EM energy at several azimuths and elevation angles. Radars that are used for precipitation estimation by operational meteorological services maneuver the antenna with a pedestal (see Figure 1.1). There are two basic modes of operating a rotating pedestal: **plan position indicator (PPI)** mode and **range height indicator (RHI)** mode. PPIs are obtained by spinning the antenna in the azimuthal direction while keeping it fixed at a constant elevation angle. A full 360 deg rotation in PPI mode constitutes a **surveillance scan** and yields a **tilt** of data. An RHI, on the other hand, keeps the azimuth fixed and varies the elevation angle. RHI mode is useful for interrogating specific storms, whereas a surveillance scan in PPI mode is more practical for operational precipitation estimation. It is advantageous to collect radar data in PPI mode at multiple elevation angles. First, the lowest elevation angle (e.g., 0.5 deg) may be blocked by terrain, buildings, or trees in some sectors. Second, it is advantageous to collect data at greater heights so as to determine additional cloud characteristics related to storm depth, severity, vertical water content, vertical ice content, hydrometeor types, microphysical processes, etc.

The movement of the pedestal on operational radars is scheduled according to antenna rotation rates and the **volume coverage pattern (VCP)**. A VCP used for precipitation measurement by the WSR-88D radar (VCP 11) is shown in Figure 1.4. In VCP 11, the antenna is rotated from 16–26 deg/sec, while it is sequentially raised in elevation up to the elevation angles shown in the figure. This VCP results in a full **volume scan** of radar data comprising 14 elevation angles completed within approximately 5 min. Despite the WSR-88D pedestal having a mechanical limitation of its minimum/maximum elevation angles of –1/60 deg, the maximum steerability of the pedestal limits the highest elevation angle used in VCP 11 to 19.5 deg (Figure 1.4). If a storm develops or moves in very close vicinity to the radar (i.e., within 10–15 km), significant

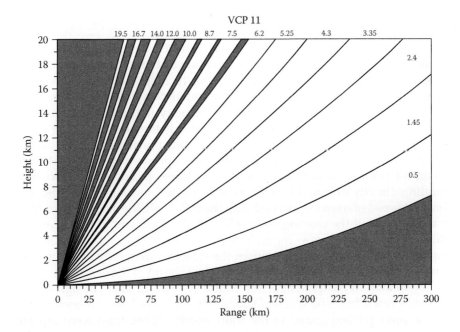

FIGURE 1.4
Radar beam heights as a function of range for 14 elevation angles comprising a volume coverage pattern (VCP). VCP 11 can be completed in 5 min and is commonly employed for quantitative precipitation estimation.

parts in the middle and upper part of the storm can go unobserved. This data void region in close vicinity to the radar is called the **cone of silence** and can result in underestimated storm top height estimates and unrealistic trends in radar severity indexes (see Howard et al. 1997). Negative elevation angles can be useful for low-altitude surveillance in valleys using radars that have been sited on mountaintops (Brown et al., 2002). Scanning at vertical incidence (90 deg), if permitted by the pedestal, can be quite useful for calibrating the radar variables.

1.3 The Radar Pulse

Operational radars transmit discrete pulses of EM energy using a **modulator** and then "listen" by discretizing the received data into range bins; this procedure is called **range gating**. The range from the radar to the target is determined as follows, considering the two-way travel of the pulse to the target and back to the receiver:

$$r = \frac{cT}{2} \tag{1.4}$$

where c is the speed of light (3×10^8 m sec^{-1}) and T in sec is the elapsed time between a transmitted pulse and the reception of the backscattered energy from the same pulse. In addition to providing the range to a target, a Doppler radar has the advantage of detecting the radial component of the target's velocity, commonly referred to as **radial velocity** (v_r) in m sec^{-1}. In other words, the Doppler velocities indicate how quickly targets are moving either toward or away from the radar. This is the same phenomenon that people observe with approaching trains. As the train approaches, the radial component of velocity toward the observer increases (unless the observer is standing directly in front of the train, which is not recommended!). So, the constant speed of sound associated with the train's whistle or engine is added to the velocity of the moving train. The radial component of the incoming train's velocity increases as the train approaches the observer who is standing safely off the tracks. This causes the effective frequency of the sound wave to increase, resulting in a higher pitched sound to the ear. As the train passes, the radial component of the train's velocity reverses in sign and is thus subtracted from the speed of sound. This causes a lower frequency and thus a lower pitched sound of the train whistle. If the train were capable of moving at the speed of sound, the observer would no longer hear the train as it passed since the effective speed of the sound wave would be zero. Just like a human ear, a Doppler radar uses a **phase detector** to measure the shift of the transmitted wave; this is the **Doppler shift** or **Doppler effect**.

Doppler velocities are more commonly used for severe weather detection, such as rotation with supercell thunderstorms, and less so for quantitative precipitation estimation.

Characteristics of the radar pulse dictate the radar data quality, resolution, sensitivity, and ambiguity of the received signals. The **pulse repetition frequency (PRF)** in sec^{-1} is the number of pulses the radar transmits per second. In the case of the WSR-88D, the PRF is approximately 1000 sec^{-1}. The reciprocal of the PRF is the **pulse repetition time (PRT)**, which is the elapsed time from the beginning of one pulse to the next one. This is approximately 1×10^{-3} sec for the WSR-88D radar. The **pulse duration (τ)** in sec is how long it takes to transmit a single pulse of energy. The **pulse length (H)** in m is the corresponding length after multiplying the pulse duration by the speed of light. PRF and the pulse length are important because they determine the maximum unambiguous range and velocities of the received signals, as well as the sensitivity and resolution of the received data.

Since the radar transmits multiple pulses at a fixed location, it can become difficult to distinguish the received signals backscattered to the radar coming from different pulses; this creates a **range ambiguity** in the received signals and can result in **range folding**, where the same echo appears at multiple ranges. The **maximum unambiguous range (R_{max})** in m for a radar is computed as follows:

$$R_{max} = \frac{c}{2PRF} \qquad (1.5)$$

where c is the speed of light (3×10^{8} m sec^{-1}) and *PRF* is in sec^{-1}. According to Equation (1.5), a greater maximum unambiguous range can be attained by reducing the number of pulses transmitted per unit time (the PRF). The major tradeoff in using a low PRF is lower quality Doppler velocity measurements. The **maximum unambiguous velocity (V_{max})** in m sec^{-1} is computed as

$$V_{max} = \frac{\lambda PRF}{4} \qquad (1.6)$$

where λ is the radar wavelength in m and *PRF* is in sec^{-1}. From Equation (1.6), the selection of a lower *PRF* results in a lower maximum unambiguous velocity. Similar to the range folding problem resulting from range ambiguities, Doppler velocities for a given target reset to 0 when the V_{max} is exceeded. In some cases, the velocities increase again until the V_{max} is reached again. This is called **velocity folding** and can be corrected to a certain degree in postprocessing of the radar images. The balance in selecting a *PRF* that yields a reasonable maximum range while maintaining quality velocity measurements is known as the **Doppler dilemma**. There are novel solutions to the Doppler dilemma such as using multiple PRFs. Additional details of these techniques can be found in Doviak and Zrnić (1985) and Tabary et al. (2006).

In addition to the PRF, which controls the R_{max} and V_{max}, a characteristic of the radar's pulse is its pulse length. The selection of the pulse length presents a tradeoff between range resolution and sensitivity. The range resolution (Δr) in m is determined as follows:

$$\Delta r = \frac{c\tau}{2} \tag{1.7}$$

where c is 3×10^8 m sec^{-1} and τ is in sec. For the WSR-88D, the range resolution (i.e., the length of the range gate in Figure 1.2) in short pulse mode ($\tau = 1.57 \times 10^{-6}$ sec) is 250 m. This corresponds to a pulse length of 500 m. Two targets must be separated by one-half the pulse length in order to distinguish them; thus the range resolution is one-half the pulse length. A longer pulse ($\tau = 4.7 \times 10^{-6}$ sec) corresponding to a range gate of 750 m is also used by the WSR-88D. The advantage of the long pulse is greater sensitivity by a factor of 3. This enables the radar to sense much weaker echoes such as those from drizzle or snow. Radars often operate in a **clear air mode** using long pulses at lower elevation angles when there are no strong echoes within the radar scanning region, also referred to as the **radar umbrella**. Thresholds based on the received backscattered energy are used to switch the radar from clear air mode into precipitation mode employing short pulses. In summary, the advantage of high-resolution range gates outweighs the cost of sensitivity loss when significant precipitation is observed by the radar.

Figure 1.2 illustrates a range gate close to the radar and then another at far range. The range resolution, dictated by the selection of the pulse length, and the beamwidth are the same for both range gates. However, the beam-spreading effect results in much different bin volumes as a function of range. The following equation can be used to approximate the bin volume (V) in m^3:

$$V = \pi \left(\frac{r\theta}{2} \right)^2 \frac{c\tau}{2} \tag{1.8}$$

where r is the range in m, θ is the beamwidth in radians, c is the speed of light (3×10^8 m sec^{-1}), and τ is the pulse length in sec. At close range, the bin volume is small and more pencil shaped, while the far-range bin resembles more of a pancake having a thickness equal to the length of the pencil. The azimuthal resolution of the radar beam depends on the rotation rate and the processing capabilities of the radar system. In the case of the WSR-88D radar, oversampling is employed in the azimuthal direction resulting in a maximum bin resolution of 250 m in range by 0.5 deg in azimuth.

In summary, the configuration of the transmitted radar pulse represents a balance between resolution and sensitivity as well as maximum unambiguous range and velocity. Shorter pulse lengths, or the duration over which the signal is transmitted, yield higher resolution data in the range coordinate. The tradeoff for range resolution is less power returned to the radar from the backscattered targets. The power loss in sensitivity is directly proportional to the increase in the range resolution. The radar PRF controls the maximum range at which echoes can be distinguished from one another at a fixed azimuth and elevation angle. Lower PRF modes of operation are generally preferable for hydrologic use of weather radar due to the extended maximum unambiguous range. However, a low PRF reduces the quality of the Doppler winds.

1.4 Signal Processing

Now that the radar has successfully transmitted a signal that encountered a target yielding backscattered energy to the receiver, the computer must process the signal to generate radar products. A radar collects a number of samples at a given range gate. The degree to which these samples are independent depend on the factors dictating the bin volume in Equation (1.7), PRF, radar wavelength, angular beamwidth, antennae rotation rate, and homogeneity of the hydrometeors being sampled. Figure 1.5 shows an illustration of individual samples at a specific range that are used to create the Doppler spectrum. The peak of the primary mode of the spectrum corresponds to the

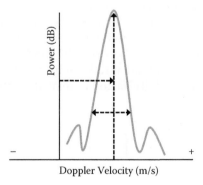

Doppler Velocity (m/s)

FIGURE 1.5
Illustration of a Doppler spectrum for a given range gate. The height of the peak of the primary mode of the spectrum is used to compute reflectivity. The shift of the peak to the right or left corresponds to the Doppler shift and is used to derive radial velocity. The breadth of the distribution is used to compute spectrum width.

average power received. This is a derivative of the Doppler spectrum and is referred to as a **radar moment**. The displacement of the peak power to the right or to the left represents the Doppler shift and corresponds to the radial component of the velocity. The width of the spectrum, or **spectrum width**, provides a quantification on the variability of the individual samples' velocities and is related to uniformity of the hydrometeors' movement within the beam. If there is a great deal of turbulence or wind shear, then individual samples will have significant variability from pulse-to-pulse and a large spectrum width will result.

The range to the signal and the amount of power received approximately dictates the **radar reflectivity factor** Z, expressed in linear units mm^6 m^{-3} as

$$Z_e = \frac{\overline{P_r} r^2}{C|K|^2} \tag{1.9}$$

where P_r is the received power in watts with the overbar representing the average of the individual samples, r is the range to the target in m, K is the dimensionless complex index of refraction of the scattering particles (equal to 0.2 for ice and 0.93 for water), and C in watts m^{-1} is the radar constant that describes the operating characteristics of the radar as follows. It can be expanded as follows:

$$C = \frac{\pi^3 c}{1024(\ln 2)} \left[\frac{P_t \tau G^2 L \theta^2}{\lambda^2} \right] \tag{1.10}$$

where π is 3.14159, c is the speed of light (3×10^8 m sec^{-1}), P_t is the peak transmitted power in watts, τ is the pulse duration in sec, G is the dimensionless antenna gain, L is the loss factor due to attenuation of the signal, θ is the beamwidth in radians, and λ is the radar wavelength in m. The gain is a measure of the antenna's ability to focus the transmitted energy relative to isotropic transmission of the signal. The radar-measured reflectivity is often referred to as the equivalent reflectivity and thus the subscripted e, while the theoretical reflectivity factor is related to the raindrop particle sizes as follows:

$$Z = \int N(D) D^6 \, dD \tag{1.11}$$

where D is the drop diameter in mm and $N(D)$, the **number concentration**, represents the number of raindrops within an interval of dD when using the discretized form of the equation. From Equation (1.11), we can see why Z when expressed in linear units is given in mm^6 m^{-3}, representing the drop diameter (typically on the order of mm) per unit volume (m^3). Reflectivity in linear units spans several orders of magnitude, so it is convenient to compress the values in units of decibels of reflectivity or dBZ:

$$dBZ = 10 \log_{10} Z \tag{1.12}$$

From Equations (1.9) and (1.11), we can see the radar-measured Z_e is related to the concentration of raindrops and their diameters, which is the primary information needed to compute the volume or mass of water within a radar-illuminated volume. This is the general basis for rainfall estimation by radar. But first, a number of assumptions must be made to get there.

Let's begin with the equivalent reflectivity measured by radar. The radar is measuring the backscattered cross-sections from precipitation within the illuminated volume. First, these particles must be Rayleigh scatterers uniformly spread throughout the volume. The amount of backscattered energy depends on the **size, shape, state,** and **concentration** of the particles. The size of the particles is relative to the polarization of the radar wave. For horizontally polarized radars, this means that the radar is sensing the horizontal component of the particles, or the drop diameters. The shape of the particles also impacts the backscattered energy. When droplets are small with diameters less than 0.5 mm, they are approximately spherical. However, as they grow, they begin to fall due to gravitational acceleration. But, frictional forces oppose gravity until equilibrium conditions are met, which corresponds to the terminal fall velocity of the drop. Like parachutists jumping out of a plane, they will first feel gravity as they begin their descent. But, they will quickly encounter these frictional forces underneath causing their clothes to flap and making a windy noise. In the case of the raindrop, this drag causes their shapes to distort and become **oblate** such that their horizontal dimension (or semimajor length a) is greater than their vertical one (semiminor length b). In this sense, as drops become larger, they begin to resemble more of a frisbee rather than a teardrop. A schematic of drop shapes for different sized diameters is shown in Figure 1.6.

The physical state, or phase, of the backscattered particles also impacts Z_e. At subfreezing temperatures (heights greater than the freezing level), ice particles are relatively small and pristine, meaning that they have not aggregated or grown at the expense of other particles (Figure 1.6). As ice particles fall and begin to melt, they become water-coated and take on the dielectric properties of water more than ice. Water has a higher dielectric constant than ice. The particles also aggregate as they melt because they essentially become sticky. This causes a sharp increase in reflectivity just below the onset of melting that is so reflective to radar it is called the radar **bright band**. As melting continues, the water-coated snowflakes become raindrops. These have smaller horizontal cross-sections D compared with the snowflakes, and they fall more quickly. Because of the higher fall speeds, they have smaller number concentrations. From Equation (1.11), both $N(D)$ and D decreases just below the melting layer and Z_e decreases.

FIGURE 1.6
Illustration of a radar scanning a stratiform cloud producing rainfall at the surface. Aloft, the radar samples pristine ice crystals with relatively low reflectivity values. As these hydrometeors fall, they encounter the melting layer. As they melt, they aggregate and become much more reflective to radar due to their water-coated surfaces. As they continue their descent, they completely melt into raindrops, which are shaped like oblate spheroids. Reflectivity decreases below the melting layer and remains approximately constant with height until the raindrops reach the surface. Note that in the presence of beam blockage, the radar is only capable of obtaining representative samples of surface rainfall in close proximity.

Equation (1.11) indicates that Z depends on $N(D)$ and D. Conventional weather radar, however, does not measure both these variables independently, but rather detects their combined effect. To deal with this, we must assume a raindrop size distribution (often referred to as the **drop size distribution, DSD**). Rain DSDs can be described with a normalized gamma distribution as

$$N(D) = N_w f(u) \left(\frac{D}{D_0} \right)^{\mu} \exp \left[-(3.67 + \mu) \frac{D}{D_0} \right] \qquad (1.13)$$

where D_0 is the equivolumetric median drop diameter in mm. N_w in mm m^{-3} is the normalized concentration defined as

$$N_w = \frac{(3.67)^4}{\pi \rho_w} \left(\frac{10^3 W}{D_0^4} \right) \qquad (1.14)$$

where the density of water ρ_w is 1 g cm^{-3} and W is the rainwater content in g m^{-3}. N_w can be interpreted as the intercept on the $N(D)$ axis of an exponential distribution having the same rainwater content as the gamma function. Next, $f(\mu)$, the shape parameter function, is defined as

$$f(\mu) = \frac{6}{(3.67)^4} \frac{(3.67+\mu)^{\mu+4}}{\Gamma(\mu+4)} \tag{1.15}$$

From Equations (1.13) to (1.15), we can see that the DSD simplifies to an exponential distribution if the shape parameter μ is assumed to be 0.

To compute rainwater contents, we must model the oblateness of the rain-drop shapes that are shown schematically in Figure 1.6. Several models relating the ratio of a drop's semiminor axis length b to the semimajor axis length a (i.e., the drop aspect ratio) to the equivolume spherical diameter D (in mm) have been proposed. Gourley et al. (2009) tested several proposed drop shape models and determined that the one described in Brandes et al. (2002) was the most correct. This models is represented as follows:

$$\frac{b}{a} = 0.9951 + 2.51 \times 10^{-2}(D) - 3.644 \times 10^{-2}(D^2) + 5.303 \times 10^{-3}(D^3)$$

$$- 2.492 \times 10^{-4}(D^4) \tag{1.16}$$

and is illustrated in Figure 1.7.

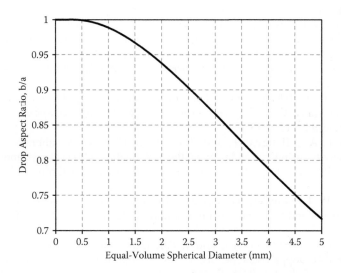

FIGURE 1.7
The dropshape model of Brandes et al. (2002) that relates raindrops' ratio of their semiminor axis length b (vertical dimension) to the semimajor axis length a (horizontal dimension) as a function of the equal-volume spherical diameter.

Note that the ratio b/a is set to unity for $D < 0.5$ mm, representing spherical shapes for very small droplets. Now, with a description of the distribution of the drop sizes and how these sizes relate to shape, we have the basic ingredients to compute a rainfall rate from reflectivity measurements.

Problem Sets

Q1: What is the Doppler dilemma? Please describe a novel solution to this issue.

Q2: What elevation angle should you use if you would like to observe a target at an altitude of 1000 m at a range of 100 km? How would your results change if the desired altitude was 100 m? Discuss your options and hardware limitations.

Q3: If in a rainfall event, droplets are of a single size and fall speed ($D = 1$ mm, $v = 5$ m sec^{-1}), and the droplet density is 15 drops m^{-3}. What is the rainfall rate in mm/hr?

━━━━━━━━━

References

Brandes, E. A., G. Zhang, and J. Vivekanandan, 2002. Experiments in rainfall estimation with a polarimetric radar in a subtropical environment. *Journal of Applied Meteorology* 41: 674–685.

Brown, R. A., V. T. Wood, and T. W. Barker, 2002. Improved detection using negative elevation angles for mountaintop WSR-88Ds: Simulation of KMSX near Missoula, Montana. *Weather and Forecasting* 17: 223–237.

Doviak, R. J., and D. S. Zrnić, 1993. *Doppler Radar and Weather Observations*. Academic Press, 458 pp.

Gourley, J. J., P. Tabary, and J. Parent-du-Chatelet, 2006. Data quality of the Meteo-France C-band polarimetric radar. *Journal of Atmospheric and Ocean Technology* 23: 1340–1356.

Gourley, J. J., A. J. Illingworth, and P. Tabary, 2009. Absolute calibration of radar reflectivity using redundancy of the polarization observations and implied constraints on drop shapes. *Journal of Atmospheric and Ocean Technology* 26: 689–703.

Howard, K. W., J. J. Gourley, and R. A. Maddox, 1997. Uncertainties in WSR-88D measurements and their impacts on monitoring life cycles. *Weather and Forecasting* 12: 167–174.

Tabary, P., F. Guibert, L. Perier, and J. Parent-du-Chatelet, 2006. An operational triple-PRT Doppler scheme for the French radar network. *Journal of Atmospheric and Ocean Technology* 23: 1645–1656.

2

Radar Quantitative Precipitation Estimation

The implementation of weather radar networks in many meteorological agencies throughout the world has changed their paradigm for severe weather monitoring and warning. It has transformed from a system of manual reporting and reacting to weather to one of automated observations and anticipating weather impacts. This latter system has resulted in a significant reduction in loss of life and property, well worth the investment of installing and maintaining the observing networks. This monitoring capability also exists for **quantitative precipitation estimation (QPE)**. However, the quantitative use of weather radar variables is the most demanding and thus requires careful processing and error considerations to convert the radar signal to a useful measurement of precipitation rates.

Prior to the advent of radar technologies, the measurement of rainfall was accomplished with in situ rain gauges. A diverse array of rain gauges exists, ranging from simple collection receptacles that observers read to weighing buckets, tipping buckets, acoustic devices, heated plates and spheres, and laser and video disdrometers that measure individual drops. Each device has its own set of advantages, costs, maintenance protocols, electricity and communication requirements, and error sources, but all in situ gauges share a common problem in representing spatial precipitation patterns. This limitation is particularly problematic with extreme rainfall and orographic precipitation, both of which are commonly characterized by strong spatial gradients. The greatest benefit of weather radar to hydrology is its potential to estimate rainfall rates at high spatiotemporal resolution (i.e., 1 km/5 min), in real time, within a radius of approximately 250 km of the radar. The following sections explain the basic procedures needed to get from reflectivity measured by conventional radar to precipitation rates with uncertainty estimates.

2.1 Radar Calibration

The calibration of radar has a major influence on the accuracy of rainfall rates. A miscalibration of only 1 dB results in bias in rainfall rates of 15%. Several options for calibrating radar are summarized in Atlas (2002). The receiver can be calibrated by injecting it with a known, transmitted signal. However, this "engineering approach" doesn't account for the combined error resulting from transmission and reception. Transmit and receive components can be jointly calibrated by using a reflective target with known scattering properties within the field of view of the radar. Such a target must be suspended or lofted, which poses challenges for large radar networks that might consist of dozens or even more than a hundred radars. Moreover, the calibration of a radar may drift in time, requiring the sphere to be positioned in front of the radar on a regular basis.

A disdrometer is an instrument that primarily measures the precipitation drop size distribution (DSD). Since it can measure the diameter of individual droplets, disdrometers also measure radar reflectivity Z (see Equation (1.10)). These in situ measurements can be compared to the radar-measured equivalent Z values in rain to identify biases due to miscalibration of the radar as in Joss et al. (1968). However, the sample volumes between a typical radar pixel aloft and a disdrometer orifice at ground differ by about eight orders of magnitude (Droegemeier et al. 2000). Furthermore, the radar samples precipitation at some height above the disdrometer (which depends on the range of the instrument from the radar, elevation differences, and beam propagation paths), so there will be a space-time lag between the measured raindrops to enter the disdrometer. This lag depends on wind velocity, the fall speed of the raindrops, and the height difference between the measurements. If this lag becomes too large, then there is a chance that microphysical processes such as melting, collision-coalescence, and drop breakup can change the character of the radar-measured Z that reaches the surface. Nonetheless, Gourley et al. (2009) used a disdrometer at 11.5 km range to evaluate the calibration of a mobile radar during the Hydrometeorological Testbed Experiment in California. Despite some scatter between the disdrometer- and radar-measured Z values, an average bias of 6.8 dB with a standard deviation of 1.3 dB was identified and later corrected to compute rainfall rate estimates. Note that the radar receiver was calibrated prior to the experiment by injecting it with a known signal. Apparently, this engineering approach was inadequate to handle the combined calibration from transmission and reception. The disdrometer approach to radar calibration is useful for individual radars, especially in a field experiment setting, but they are often not feasible for calibrating large radar networks.

Spatial maps of radar-estimated precipitation are often computed using data from several radars. Even if the radars are calibrated within 0.5 dB of each other, lines of data discontinuities or **radar artifacts** often arise where a precipitation estimation algorithm switches using data from one radar to the neighboring one. These artifacts are most noticeable for long-term accumulations of rainfall such as daily accumulations. This problem can be dealt with in the precipitation algorithm by spatially interpolating or smoothing data between neighboring radars. Another approach is to compare the Z values in rain from neighboring radars at these equidistant lines to identify relative calibration differences. This approach of comparing remote-sensing observations to one another, as opposed to in situ data, has been shown to be quite useful for calibrating radar networks (Gourley et al. 2003). Bolen and Chandrasekar (2000) used radar Z measured from space and compared those to NEXRAD radars to identify miscalibrated radars. The advantages of this approach are the comparisons between measurements from similar radar bin volumes, and there are generally very large numbers of matched data pairs. These comparisons, however, reveal relative calibration differences rather than unambiguously identifying which radar is miscalibrated. It is possible to complement the relative Z differences from a ground network of radars to Z measured by spaceborne radar aboard the Tropical Rainfall Measurement Mission (TRMM) and Global Precipitation Measurement (GPM) mission. The spaceborne radar may not be perfectly calibrated itself, but it is stable and traverses regions covered by many different radars. By comparing Z values in rain among all ground radars to their neighbors and also to a stable calibrator from space, it is possible to "level the field" so that all radars are well calibrated.

2.2 Quality Control

Now that the Z data have been bias-corrected for radar miscalibration, which may require large samples of comparisons over hours or even days of precipitation, every single bin of radar data must be carefully scrutinized to remove deleterious effects from nonmeteorological scatterers on the ground, biota in the atmosphere, planes, chaff, etc. Remember that radar was originally developed to detect planes, ships, and submarines on the ocean surface, so it comes as no surprise that weather radar sees many nonweather targets. Some researchers have devoted a great deal of their careers to the single subject of ground clutter removal from radar, which is a testament to the difficulty in accomplishing this task. It would be impractical to provide a comprehensive review of every algorithm to remove nonweather echoes, but this section covers the basic approaches.

2.2.1 Signal Processing

The first level of screening takes place at the spectral level, which is prior to the stage at which reflectivity is estimated. Radar actually measures several independent samples within a given range bin. If the samples are associated with no Doppler shift, i.e., a radial velocity of 0 m sec^{-1}, then this indicates a stationary target, probably from the ground or a building. Hydrometeors have nonzero Doppler velocities because they fall and are displaced horizontally with the wind. At this point, it might seem rather trivial to just remove all echoes associated with no Doppler shift. However, even when the wind is prevailing from a uniform direction, there will be a line where the radial velocity shifts from negative to positive; this is called the **zero isodop**. Consider a southerly wind. There will be negative radial velocities to the south of the radar (moving toward the radar), positive to the north (moving away from the radar), and zero velocities due east and west. If a filter were to remove all echoes with no Doppler shift, then echoes to the east and west of the radar along the zero isodop would be mistakenly removed.

Conversely, there are numerous nonmeteorological targets that are nonstationary. All biota in the atmosphere such as birds, bats, and insects are generally in flight and are affected by the prevailing wind. Even some ground targets like trees and windmills have nonzero velocities. The leaves of trees flutter in the wind, and the tree trunk itself gently sways. Even this small motion is detectable by radar. The recent development of wind farms has resulted in a great deal of ground targets that have moving parts. These are often placed on exposed ridges and are thus within the view of nearby radars. These kinds of targets pose significant challenges to quality control procedures. Since each and every bin around a radar must be scrutinized, automated algorithms must be developed, tuned, and implemented to screen out nonmeteorological echoes. The first indication of ground clutter is echoes with zero Doppler velocity in regions away from the zero isodop. These echoes are first evident at the spectral level of processing, and are subsequently removed.

2.2.2 Fuzzy Logic

Algorithms must be employed to discriminate nonmeteorological from meteorological echoes. These algorithms range in complexity from simple thresholds placed on variables, thresholds applied to multiple variables as in decision tree logic, fuzzy logic, neural networks, and combinations therein. Fuzzy logic algorithms are well suited for radar observations because they incorporate information from multiple variables with different weights and are less susceptible to misclassifications due to noise in the measured variables. This section covers the general design of a fuzzy logic algorithm. Additional variables from radars with polarimetric capability are presented in Chapter 3.

A fuzzy logic algorithm accommodates imperfect measurements for decisions that must be made about a process that is not perfectly understood or explained by measurable data. Nonmeteorological scatterers have several identifiable radar characteristics, such as zero velocity, but no single variable unequivocally discriminates meteorological from nonmeteorological scattering. Fuzzy logic algorithms consider multiple radar observations; thus the impact of a single, perhaps noisy variable is minimized. The key to the success of a fuzzy logic algorithm lies in the designation of its **membership functions**. Membership functions utilize qualitative and/or quantitative information from observations, theory, or simulations. They describe the range of values a variable possesses to aid in the decision.

The membership functions of Zrnić et al. (2001) have trapezoidal shapes, while those described in Liu and Chandrasekar (2000) have continuously differentiable beta functions. Gourley et al. (2007) derived membership functions using Gaussian kernel density estimation (Silverman 1986) that are based entirely on radar observations. First, cases comprised of several hours of observations are identified in which there was no precipitation but evident ground clutter from nearby buildings, trees, etc. Then, a function describing the typical ranges of values for each radar variable in ground clutter is computed as follows:

$$\hat{f}(x) = \frac{1}{\sigma\sqrt{2\pi}} \sum_{i=1}^{n} e^{-\left[(1/2)\left(\frac{X_i-x}{\sigma}\right)^2\right]}, \tag{2.1}$$

where $\hat{f}(x)$ is the density function, σ is the smoothing parameter or bandwidth, n is the total number of data points, X_i is the ith observation of the variable, and x is the independent variable. The bandwidth controls the smoothness of the function, and it can be estimated using Silverman's rule ($\sigma = 1.06SDn^{-1/5}$) where SD refers to the standard deviation of the raw data distribution. Applying a Gaussian kernel is equivalent to drawing a Gaussian curve on top of each data point, and then adding up all the individual functions using the principle of superposition.

Figure 2.1 illustrates how individual Gaussian curves (in gray) are produced for each individual data point (denoted with + signs). The process is repeated for all of the data points, and after the curves are added and normalized, the density function $\hat{f}(x)$ (curve in black) is produced. The density function can be nonlinear, it can have multiple modes, and it is

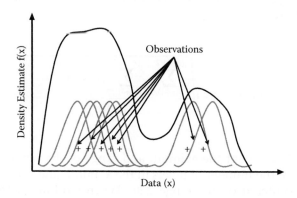

FIGURE 2.1
Illustration of kernel density estimation using Gaussian curves. A Gaussian curve is drawn for each observational data point along the x-axis. Each of the curves are added using liner superposition to arrive at the final curve that gives an estimate of the data density.

continuously differentiable. It can be thought of simply as a smoothed fit to a histogram of the data. For fuzzy logic to function properly, density functions need to be produced for the radar variables for all classes. A **class** is the predictand, or what the algorithm is designed to identify. For discriminating hydrometeors from nonhydrometeors, cases are selected for situations with pure precipitation (no nonmeteorological scatterers) and with nonmeteorological scatterers alone.

In addition to incorporating the standard variables measured by radar in a fuzzy logic algorithm, it is also useful to consider their temporal and spatial derivatives; this is because radar observations of nonmeteorological scatterers can be discontinuous and noisy in space but consistent in time. Consider the theoretical Z that would come from a building versus that of an individual convective storm. The building will be very reflective, perhaps similar to a strong thunderstorm cell, but these high values will be quite consistent in time (the building is not moving). Furthermore, bins just adjacent to the building will likely have very low values of Z. The thunderstorm, on the other hand, will exhibit higher temporal variability in Z because hydrometeors are rapidly ascending, descending, or being advecting laterally. Furthermore, a strong thunderstorm typically extends horizontally at least a few kilometers and often can extend tens of kilometers. A sigma variable measures the temporal consistency of reflectivity by computing the mean absolute difference in Z between adjacent pulses at a given range gate (Nicol et al. 2003). Low variability (<5 dB) is more typically associated with stationary targets such as ground clutter. Opposite of temporal stability, very high spatial variability of radar variables can be indicative of nonmeteorological scattering. A useful derivative to compute to assess spatial variability of a radar variable is the root-mean-squared difference, or more commonly, the **texture**. Texture fields assess the spatial variability of a given variable (y) within a user-selectable ($m \times n$) window and are computed as follows:

$$Texture\left(y_{a,b}\right) = \sqrt{\frac{\displaystyle\sum_{i=-(m-1)/2}^{(m-1)/2} \sum_{j=-(n-1)/2}^{(n-1)/2} \left(y_{a,b} - y_{a+i,b+j}\right)^2}{(m)(n)}}, \tag{2.2}$$

where a and b represent the azimuth and range of the range gate; and m and n represent the number of pixels in the azimuth and range directions, respectively, centered on the range gate. When working with raw radar data, it is important to recognize that the bin volumes increase with range. Thus, the texture variables will naturally increase in precipitation with range merely due to the much larger region over which the spatial variability is assessed.

The next step in fuzzy logic is to compute the **aggregation value (Q)** at each range bin. The aggregation value for a given class represents the strength in identifying it given the combined use of the radar variables and their derivatives. It is defined as follows:

$$Q_i = \frac{\sum_{j=1}^{n} \hat{f}(x)_j \times W_j}{\sum_{j=1}^{n} W_j} \tag{2.3}$$

where the summation is performed for each ith class using each of the jth variables and derivative fields. Some fuzzy logic algorithms use multiplication rather than aggregation shown in Equation (2.3). The danger in using a multiplicative approach is that a membership value of zero from a single membership function for a given class will prohibit that class from being assigned, even if the measured variable was in error or was noisy. Thus, the aggregation approach is more useful to remote sensing observations. The weighting applied to each radar variable, W, can be determined in an objective way as originally proposed by Cho et al. (2006).

Figure 2.2 illustrates two hypothetical density functions: one quantifies the data distribution for a given radar variable for nonhydrometeors, and the other is for hydrometeors. The weighting applied to this particular variable is computed from the degree of overlap between the two functions and is inversely proportional to the area shaded in gray. So, if two functions overlap perfectly, meaning the radar variables have the same values in both

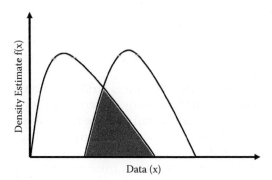

FIGURE 2.2
Two density functions with overlapping regions highlighted in gray. These illustrate how radar observations for two different classes have different distributions, but there is overlap between them. The inverse of the area of overlap is used in a fuzzy logic scheme to quantify the weight this particular radar variable will have. If there is a small area of overlap, then this indicates the variable is a good discriminator for that class, and a great deal of weight is supplied to the radar variable as a result.

precipitation and ground clutter, then the weight applied to this variable will be zero. On the other hand, very little overlap between the functions indicates that the radar variable is a good discriminator by itself and receives a great deal of weight. If there is no overlap between functions, then the weight becomes infinite. This may appear as problematic for the fuzzy logic formulation, but if there is no overlap between functions, then the variable is a perfect discriminator alone and a threshold placed on the variable, as in a decision tree logic, will suffice in the discrimination.

The next step in fuzzy logic is to determine the maximum Q_i value (for each class). It is possible at this point to compute a strength of classification or an uncertainty estimate by comparing the values for Q_i. Q values that are close to one another indicate that the strength of classification is weak and the final class is associated with high uncertainty. The final step in fuzzy logic often applies a despeckling algorithm to create spatially consistent fields. For example, neighboring classifications are examined around a given pixel. In precipitation, it is very unusual to have a single pixel identified as precipitation surrounded by nonmeteorological scatterers. So, the despeckling filter examines this pixel and then reassigns it to the nonmeteorological class due to its isolated, nonphysically realistic behavior.

2.3 Precipitation Rate Estimation

Calibrated reflectivity values describing the size, shape, state, and concentration of the hydrometeors within the radar sampling volume are used next to compute precipitation rate. If there are no beam blockages in the vicinity of the radar, then reflectivity measured at the lowest elevation angle, 0.5 deg for NEXRAD, should be used for QPE. In regions with complex terrain, bins are selected at multiple elevation angles to construct the **hybrid scan** for QPE.

Figure 2.3 shows how the hybrid scan is built based on two rules: (1) the center of the beam must clear the underlying terrain by more than 50 m, and (2) the beam blockage between the bin and the radar at the same elevation angle must be less than 60%. So, we can see that having a tall mountain close to a radar can result in unusable data at low elevation angles for all range bins beyond the obstruction. Data from range bins that have partial blockages less than 60% from terrain at closer ranges are still used, but 1 dB is added to each measured Z value per 10% blockage (2 dB added for 20% blockage and so on). Note that the blockages are computed using a digital elevation model (DEM). These DEMs do not incorporate anthropogenic factors such as towers and structures, nor do they include vegetation canopy. The heights of trees in the vicinity of a radar can be a very significant blockage requiring manual correction of the hybrid scan. Once the

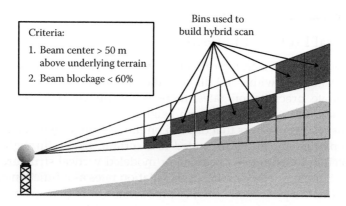

FIGURE 2.3
Construction of the hybrid scan using multiple elevation angles over complex terrain.

hybrid scan lookup table is established, it is generally fixed for a given site unless there are significant changes with buildings, towers, or trees near the radar. After the DEM-based hybrid scan is constructed, it is good practice to examine long-term accumulations during widespread, stratiform rain situations. The resulting accumulation maps will reveal artifacts, such as discontinuities evident in the azimuthal direction, that are due to blockages. Manual editing of the hybrid scan lookup table can alleviate many of these low-level blockages.

Reflectivity from the hybrid scan is used to compute two-dimensional fields of precipitation rate in spherical coordinates (range, azimuth). The general form of reflectivity-to-rainfall relationships, or Z–R equations, is a power-law as

$$Z = aR^b \tag{2.4}$$

where a is the prefactor and b is the exponent. The two most common Z–R relations are the NEXRAD default for convection ($a = 300$, $b = 1.4$) and the Marshall–Palmer relation ($a = 200$, $b = 1.6$) generally applied to stratiform rain. The Marshall–Palmer relation (Marshall and Palmer 1948) comes from an exponential DSD with fixed slope and intercept parameters. In Equation (2.4), Z is in linear units ($mm^6\ m^{-3}$) and R is in mm hr^{-1}. The same power-law relation is used to compute snowfall rates. With radar, the snow water equivalent (SWE) is the variable that is estimated rather than the snow depth. The latter variable depends on the snow density, which varies with temperature and moisture. In fact, temperature may also affect the Z–S parameters for SWE estimates. Myriad values of a and b parameters have been reported for rainfall and SWE estimation specific to geographic regions, seasons, storm lifecycles, hydrometeor types, DSDs, etc. Errors in the Z–R relation are discussed in Section 2.7.

2.4 Vertical Profile of Reflectivity

Assuming all nonmeteorological echoes have been successfully removed, Z is well calibrated within 1.0 dB, and the parameters of the Z–R relation yield accurate precipitation rates, there is the unavoidable circumstance that the radar bin volume and beam heights increase with range from the radar. This results in errors that depend on range. **Range-dependent errors**, biases in particular, can be mitigated to a certain degree through the use of a correction based on an observed or modeled vertical structure of precipitation. Simpler methods adjust precipitation rates as a function of range. However, this simplification becomes problematic in regions where there is beam blockage and the hybrid scan is composed of different elevation angles that vary with azimuth. The range to a target may be the same between different azimuths, but the beam heights are drastically different.

It is better practice to implement a model describing storms' **vertical profile of reflectivity (VPR)** to quantify the variability of reflectivity with height. This VPR model can be used to correct for sampling height- and range-dependent errors, namely biases. This permits Z measured by radar at a given height and range to be adjusted so that it better represents what is occurring at the surface. A number of methods have been proposed to identify the VPR including reconstruction from volumetric radar data (Andrieu and Creutin 1995; Andrieu et al. 1995; Vignal et al. 1999; Borga et al. 2000; Seo et al. 2000; Germann and Joss 2002; Kirstetter et al. 2010) and by describing or modeling the VPR with parameters (Kitchen et al. 1994; Tabary 2007; Matrosov et al. 2007; Kirstetter et al. 2013a).

This section presents a general approach that illustrates the correction procedure. Figure 2.4 shows general VPR models for (a) convection and (b) stratiform precipitation, the most elementary level of storm segregation. Convective storms have greater updraft velocities (on the order of 10 m sec^{-1} that can be quite vigorous in supercells, reaching 50 m sec^{-1}) and tend to occur over land more frequently during the warm season. These storms reach greater heights in the troposphere, often reaching the tropopause, are often associated with electrical activity, and produce more intense rainfall rates.

Stratiform rainfall systems, on the other hand, are more common over land during the cool season and also trail convection in mesoscale convective systems. Stratiform rain has lower updraft velocities (up to 1 m sec^{-1}), tends to be more widespread, and results in weaker rainfall (and snowfall) intensities. The convective VPR in Figure 2.4 illustrates that the profile is more upright, meaning that the reflectivity does not vary greatly with height. This means that a reflectivity sample taken aloft (e.g., 5 km) is approximately representative of rainfall rates experienced at the surface. In many cases, VPR correction for convective storms leads to slight corrections that can be overshadowed by other uncertainties such as the conversion of Z to R.

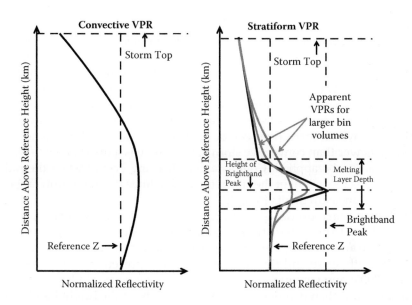

FIGURE 2.4

Model vertical profiles of reflectivity for convective and stratiform echoes. These models can be designated by the parameters listed in text. The gray curves correspond to radar-sampled VPRs at longer ranges from radar, illustrating the smoothing effect caused by beam broadening.

VPR corrections for stratiform rainfall, on the other hand, are much more germane. Figure 2.4 shows how reflectivity has little variability with height in the rain region. Then, it increases significantly in the bright band. Rainfall estimates have been shown to be overestimated up to a factor of 10 when using uncorrected measurements taken within the bright band (Smith 1986). Above this layer, reflectivity decreases again in the pristine ice region. The VPR model in Figure 2.4 describes the vertical variability of reflectivity in stratiform using the following parameters: (1) depth of the melting layer, (2) height of the bright band peak, (3) bright band peak value, (4) decrease in reflectivity from the height of the bright band peak to the top of the melting layer, (5) decrease in reflectivity from the top of the melting layer to the storm top height, and (6) storm top height. These parameters can be estimated using volumetric radar data, climatological radar data, and even from upper air data from numerical weather prediction (NWP) analyses and from radiosonde observations. Once the vertical structure is modeled, the next step is to account for beam spreading. The impact of the bin volumes increasing with range causes the radar to observe a smoothed version of the actual VPR. Two of these apparent VPRs are shown as gray curves in Figure 2.4. The smoothing, or averaging of data in the vertical, increases as the bin volumes increase with range. The apparent profiles can be estimated using Equation (1.7) superimposed on the VPR model.

The next step is to correct the radar-observed reflectivity measured at some height so that it represents the reference reflectivity value at or very near the surface. The method uses the apparent VPR, which depends on the bin volume at that range and then estimates the beam center height from Equation (1.3). These factors and the VPR model determine the amount of reduction needed for measurements taken near the melting layer and increases needed for measurements taken above it. There is generally a vertical limit to which corrections can be applied. Stratiform systems tend to be much shallower than convective storms, thus there is a finite range to which corrections can be applied. Complete overshooting occurs when the radar beam exceeds the height of the storm top and defines the maximum range of useful radar measurements for QPE.

VPR corrections are designed to reduce range-dependent biases, which are considered to be systematic. Random errors result from unrepresentative VPRs. There are numerous meteorological scenarios than can result in VPRs that vary with space and time and are thus unrepresentative. Gourley and Calvert (2003) and Giangrande et al. (2008) developed bright band retrieval algorithms and then demonstrated that the height of the bright band could change by as much as 2 km in altitude within 6 hrs. Spatial variability of the melting layer can be significant in the presence of strong temperature gradients, such as near a cold front. The melting layer also descends due to adiabatic cooling from orographically induced updrafts and diabatic cooling from melting; thus we expect lower melting layer heights on windward slopes of mountain chains (Minder et al. 2011). This VPR representation error can be magnified by increased precipitation rates from low-level orographic enhancement that may not be observed by radar. Thus, orographic effects impact the spatial variability of the VPR as well as the magnitude of reflectivity increase at low levels. Nonetheless, VPR correction methods are useful to reduce range-dependent biases and are thus a recommended practice for estimating rainfall in stratiform precipitation systems.

2.5 Rain Gauge Adjustment

Rain gauges provide useful measurements of rainfall at a point. For this reason, they are essential for evaluating and improving radar-based QPE algorithms, despite their significantly large sampling volume differences from radar bins. There are many different types of rain gauges ranging from weighing buckets to tipping buckets, heated plates, and laser and video disdrometers. It has been noted that they have their own set of errors (Zawadzki 1975; Wilson and Brandes 1979; Marsalek 1981; Legates and DeLiberty 1993; Nystuen 1999; Ciach 2003). If they are not shielded from wind, then the instrument can perturb the wind field, resulting in undercatch.

Other errors relate to the specifics of the instrument itself such as splash-out with heavy rainfall and lack of sensitivity to very light rainfall rates (i.e., not enough rainfall to tip the bucket and register rainfall). These instruments must also be regularly calibrated and maintained, which is particularly necessary for sites where lots of plants, birds, and insects congregate around the instruments. But the largest limitation by far is their inability to adequately represent the spatial and temporal variability of rainfall fields except over very dense networks covering small domains. This is the greatest advantage of radar measurements.

At a given point, rain gauges are generally considered to be of higher quality than radar-based estimates of precipitation. They are thus useful for evaluating radar QPEs and for correcting them in near real time. The two general approaches for correcting QPEs using rain gauges are (1) mean field bias correction and (2) spatially variable bias correction. Mean field bias correction simply matches each hourly rain gauge estimate to collocated radar bins (or neighborhood of bins). The radar QPEs are summed for the hour and compared to the sum of the rain gauge accumulations. A $\Sigma(G)/\Sigma(R)$ ratio is computed, which assumes the radar errors are multiplicative, and multiplied back to the original radar QPEs. This method maintains the spatial variability resolved by the radar QPEs and then adjusts them on an hourly or even daily basis by removing their mean field bias. These mean field bias corrections are effective for radar-based QPE errors that are spatially homogeneous. The only error that properly falls into this category is miscalibration of the radar signal, which affects all bins around the radar equally. Most other errors have nonnegligible spatial variability.

Forecasters at the Arkansas–Red Basin River Forecast Center originally developed a technique of spatially variable bias correction called P1. This effort was largely in response to the installation of the Oklahoma Mesonet, which provides standard surface weather observations including rain gauges in every county in the state. The method works by comparing radar QPEs at gauge locations and computing the bias. In this case, the bias is specific to each gauge location and not lumped together as in the mean field bias correction. Then, the bias field is spatially interpolated using a method such as regression, inverse distance weighting (IDW), or kriging. Then, the smoothed, spatially variable bias field is applied back to the original radar QPE field. The advantage of this approach is that it considers spatially heterogeneous errors that are common with radar QPE such as $Z–R$ variability, nonweather scatterers, and range-dependent biases. For the adjustment to be effective, there must be a reasonably dense gauge network surrounding the radar to resolve the nonhomogeneous errors. Furthermore, the method is sensitive to errors in individual gauge measurements and can end up corrupting the radar-based QPEs. The mean field bias correction method is less susceptible to individual errors due to the aggregation of all the gauge accumulations. The gauge data must be very carefully quality controlled for the spatially variable bias correction method to be effective. This is a challenging prospect since rain gauge networks tend to be operated by myriad different organizations ranging from national meteorological services, local municipalities, hydroelectric power companies, state climate surveys, etc.; thus, they are subjected to inconsistent levels of maintenance and quality control. Radar networks, on the other hand, are frequently operated and maintained nationally by the government and have much more standard operating and error characteristics.

2.6 Space-Time Aggregation

Radars can measure rainfall rates on their native spherical grid with bin resolution depending on the operating characteristics of the radar. This resolution is nominally 1 deg in azimuth by 1 km in range and estimates are computed on a 5 min basis. These rainfall rates must be summed typically to hourly accumulations in order to be adjusted by rain gauges. Additional aggregation takes place to compute 3-, 6-, 12-, and 24-hour, 72-hour, 10-day, monthly, etc., accumulations. Some gauge networks provide daily accumulations such as manual observer networks. These data can be introduced to modify the 24-hourly accumulations.

Rainfall estimates from neighboring radars comprising a network can be combined or mosaicked to create rainfall maps covering a larger, national spatial domain. The mosaicking methods differ in their complexity. Because the native radar coordinates are centered around each radar, it is more practical to resample the rainfall estimates onto a common Cartesian grid; this accommodates the mosaicking of data from adjacent radars onto a single grid. The simplest mosaicking approach merely chooses the rainfall estimate from the radar closest to each grid point. This method is notorious for creating linear discontinuities in the rainfall fields at points equidistant (at the same range) from neighboring radars. These mosaicking artifacts indicate that one or both of the radars have bias in their rainfall estimates.

Examples of this phenomenon are illustrated in Figure 2.5. This is a 10-day accumulation of estimated rainfall from the Mobile, Alabama (KMOB); Elgin Air Force Base, Florida (KEVX); and Tallahassee, Florida (KTLH), NEXRAD radars. The linear discontinuities are circled in black. They indicate that rainfall estimates from the KEVX radar are higher than from the neighboring radars. Rainfall estimates from the algorithm running on the KEVX data are approximately 5 in (127 mm) or 50% greater than from KMOB. To the east of KEVX, the differences are on the order of 2.5 in (63.5 mm) or about double the rainfall estimates from the KTLH radar. Clearly, we can see the large uncertainties that can result in radar-based rainfall estimation and the artifacts that are caused by data mosaicking.

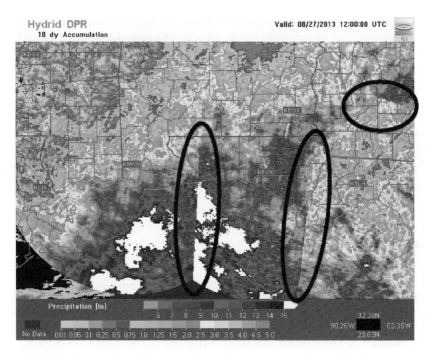

FIGURE 2.5
Linear discontinuities in a 10-day accumulation of rainfall from mosaicked radar data. The accumulation period ends at 1200 UTC on August 27, 2013.

These discontinuities can result from calibration differences between the radars, different equations used to convert Z to R, different parameters in the VPR models used to reduce range-dependent biases, partial beam blockage, power losses due to attenuation, and different beam center heights. Some of the errors caused by radar miscalibration, Z–R, VPR parameters, partial blockages, and attenuation are largely algorithmic and can be corrected or minimized using software solutions. Discrepancies in beam center heights can result from differences in beam propagation paths. Small-scale atmospheric conditions dictate the beam propagation paths. Although propagation paths can be modeled using profiles of atmospheric pressure, humidity, and temperature, some vertical gradients can be unresolvable by upper air observing systems. Large uncertainties in the beam center heights result and can cause radars to sample at different heights even though the precipitation being sampled is at the same equidistant range from both radars. This effect will cause the discontinuities. Radars most susceptible to this mosaicking artifact are those near oceans that experience a marine boundary layer and adjacent radars with different elevations.

Alternative mosaicking strategies have been developed to mitigate the visual artifacts caused by blending data from neighboring radars. Tabary (2007) proposed a mosaicking scheme that weights precipitation estimates from adjacent radars based on their quality as judged by their influence by ground clutter, degree of beam blockage, and altitude of the beam. Once the weights are empirically estimated for a given grid point, the mosaicking scheme selects the rainfall estimate from the radar that has the highest weight, which is equivalent to using the estimate with the best quality. Zhang et al. (2005) developed a scheme that first interpolates reflectivity data from spherical coordinates to a common 3-D Cartesian grid using objective analysis. Data from adjacent radars are then mosaicked based on a weighted mean, where the weight is based on the distance between an individual grid cell and the radar location. This method is more computationally expensive, but it is effective in interpolating radar data and minimizing artifacts.

2.7 Remaining Challenges

Despite all the processing steps and correction procedures applied to radar precipitation estimates, challenges still exist. In terms of data quality, not all nonweather scatterers are completely removed from radar QPE accumulations. Scattering from biological targets like birds and insects is difficult to discern from hydrometeors because these have similar sizes and velocity signatures as raindrops. Fixed structures like wind farms with moving blades can also be problematic for radar rainfall estimation. If the parameters of the data quality algorithms are tuned to remove all these nonweather scatterers, then actual precipitation can get accidentally removed.

Radar coverage at low levels (e.g., below 3 km above ground level) can be a real limiting factor in the quality of radar QPE (Figure 2.6). Maddox et al. (2002) modeled the effective radar coverage by NEXRAD over the conterminous United States. They showed very few gaps in radar coverage east of the Rocky Mountains with a few exceptions in the Rio Grande Valley of Texas and some mountainous locations in the Appalachians and Quachitas of southeast Oklahoma. Low altitudes (3 km AGL) of the Intermountain West, which is roughly defined as the region east of the Sierras and west of the Rocky Mountains, is approximately 50% covered by the NEXRAD network. This means that even following VPR correction, there will be large random errors. And, in many regions the radars completely overshoot the precipitating systems and thus there is no signal to correct. This situation is especially problematic during the cool season when the storms do not have great vertical extents as they do during the warm season. The radar coverage problem can be improved only by introducing observations from additional platforms. These can be gap-filling radars, instruments aboard satellites, gauges, and even precipitation analyses from numerical weather prediction models.

FIGURE 2.6
Radar coverage by the NEXRAD network for altitudes below 3 km above ground level. The colors correspond to multiple radars overlapping a given region.

The use of a single, deterministic relationship to convert Z to R assumes a unique DSD that describes what the radar is sampling. Variability in the DSD as a function of storm type, geographic location, season, storm lifecycle, updraft region of storms, etc. has been documented in numerous studies. Some approaches address DSD variability with conventional radars such as identifying different precipitation types and applying different $Z-R$ equations. A contemporary algorithm that conducts automated precipitation typing, adaptive $Z-R$ selection, and rainfall estimation is detailed in Chapter 4. Compared with the breadth of studies that have been conducted on estimating rainfall from radar, very little has been done with SWE. In situ observations of SWE have their challenges due to disruption of the wind field by the collecting gauges and spatial representation limitations. Snow accumulations are strongly influenced by the underlying terrain and local wind patterns that are organized on very small scales. The lack of large samples of accurate SWE accumulations by in situ instruments has been a limitation in the development and evaluation of radar-based snow algorithms. The advent of dual-polarization radar technology offers great potential to improve the quality of radar variables and to retrieve precipitation rates conditioned on different DSDs.

2.8 Uncertainty Estimation

Using deterministic radar QPEs without considering their underlying distribution assumes that they are error free or that their uncertainties are negligible in the application for which they are being used. For the latter, consider a coupled modeling system that uses radar QPEs as forcing to forecast the advection, dispersion, and reaction of a contaminant in a river. In this case, the uncertainties in the radar-estimated rainfall might be overshadowed by the initialization of the location and extent of the chemical spill, uncertainties about the specific chemical that was released, modeling of its transport underground, potential reaction with other, unknown chemical species in the groundwater and river water, etc. But, in general, it is wise practice to estimate and utilize uncertainties associated with radar QPE.

A number of radar rainfall error models have been proposed. An error model is the first step toward computing the rainfall uncertainty associated with a deterministic QPE, probabilistic QPE (PQPE), and generating ensembles of QPE. A brief review of proposed error modeling approaches is provided here; readers are referred to Mandapaka and Germann (2010) for a complete review. The first error modeling approach attempts to describe all the individual errors specific to a given radar's hardware, operating characteristics controlling the resolution and sensitivity of the estimated rainfall fields, and software used to estimate rainfall. This approach of quantifying individual errors and then superimposing them to yield a total error for the rainfall field requires detailed knowledge of the individual errors that may be specific to the radar system, how each radar is operated, rainfall regime, data quality control, assumed DSD, VPR correction methodology, and $Z–R$ equations. While these methods have a strong theoretical basis, it can be a challenging prospect to quantify all the individual errors, their interactions, and their propagation for a large radar network such as NEXRAD. Moreover, the resulting error model may not apply to other radar systems outside the network.

A more general approach in error modeling involves correcting the radar QPEs to the best extent possible using the methods outlined in this chapter and then modeling the remaining error, or the residual, using an independent reference from rain gauges. Recall that rain gauges are often used in real-time algorithms to adjust radar rainfall estimates as discussed in Section 2.5. These methods must ensure that the reference used to define the residuals is independent. This can be accomplished by withholding some gauges from the real-time estimates or by employing values from an independent rain gauge network that might not have been available for real-time use. QPE error models decompose the residuals into biases describing the systematic (unchanging) errors of the radar QPEs, conditional biases such as dependence on radar sampling height, season, rainfall regime, rainfall rate, space-time resolution of rainfall rate estimates, etc., and then the random errors (Ciach et al. 2007). After the biases are eliminated, the statistical properties of the residual errors are evaluated through examination of their space-time structure, typically evaluated with second-order moments such as correlation. Villarini and Krajewski (2010), Germann et al. (2009), and Habib et al. (2008) all showed that log-transformed random errors followed a Gaussian distribution and were correlated in space and time. Kirstetter et al. (2010) quantified the space-time correlation of residual errors using a semi-variogram. This geostatistical visual aid is created by computing the residual errors, purposefully perturbing them them in space and time (called **lagging**), and then computing their covariance. The resulting plots summarize the spatiotemporal dependence of the residual errors.

A recent method for PQPE was proposed by Kirstetter et al. (2014) to quantify the radar rainfall uncertainty at the measurement scale of 1 km/5 min. *These are the scales that must be considered when monitoring intense rainfall associated to flash flooding.* Moreover, PQPE at such fine scales can be used to comprehensively evaluate rainfall estimates from instruments aboard low Earth orbiting platforms. Satellite-based passive and active microwave sensors essentially observe a snapshot of precipitating systems as they orbit the Earth. The aforementioned gauge-based approaches do not readily accommodate these high-resolution applications because the gauges generally do not provide accurate instantaneous rainfall rates, but rather accumulations at 15 min, hourly, or daily time scales. The new method creates a reference rainfall dataset following the method discussed in Kirstetter et al. (2012, 2013b). Hourly, spatially variable biases computed from rain gauges are applied downscale to radar-estimated rainfall rate fields at 1 km/5 min resolution. This method assumes that hourly biases do not have significant temporal variability at subhourly scales. Next, the datasets are partitioned according to their precipitation type: convective, stratiform, tropical, bright band, hail, and snow. Details on the logic involved in the precipitation type segregation are provided in Chapter 4. For now, we can assume that the different precipitation types are associated with different microphysical processes and drop size distributions, impacting the parameters in the $Z–R$ relation. The datasets are filtered to remove reference rainfall rates that were too heavily bias-corrected by hourly rain gauges (i.e., 0.01 < correction factors < 100) and the precipitation type must have been consistent at the candidate pixel throughout the hour.

Data points (indicated by x in Figure 2.7) of the filtered reference rainfall rate, R (mm hr^{-1}) are plotted as a function of Z (dBZ) for each precipitation type. The data points suggest a power-law relationship between Z and R, which is typically assumed. The next step involves fitting an error model to the observed values of Z and R. The generalized additive model for location, scale, and shape (GAMLSS; Rigby and Stasinopoulos 2005) technique is used to create the smooth curves shown in Figure 2.7. These can be considered as empirical fits to the data points that now describe the distribution of R for a given radar observation of Z and precipitation type. Figure 2.7 shows the median of the distribution, or the 50% quantile denoted as Q_{50}, as well as the

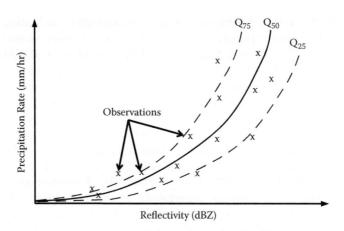

FIGURE 2.7
Illustration of method for computing probabilistic QPE at 1 km/5 min scale based on gauge-corrected radar observations and radar reflectivity factor.

25% and 75% quantiles (Q_{25}, Q_{75}). The distance between these latter two quantiles can be used to estimate the uncertainty for a rainfall rate associated with a given measurement of Z and precipitation type. The error model describes the entire data distribution and can be used to accommodate the intended application of radar rainfall estimates. For instance, in the event that the user only requires deterministic values of radar-estimated rainfall rates (such as when they are assumed to be overshadowed by other uncertainties in the modeling system), then the GAMLSS model would provide the rainfall rates associated to the Q_{50} curve. If an application was particularly sensitive to outliers, such as threshold rainfall rates to trigger flood alerts, then the user might require the upper tails of the distribution from 95% or 99% quantiles.

Now that the radar-estimated rainfall rates have been modeled using the GAMLSS technique as a function of Z and precipitation typology, a number of useful products can be derived. The relative uncertainty of rainfall estimates is readily computed using the difference between two quantiles normalized by the median precipitation rate (e.g., $(Q_{75}-Q_{25})*100/Q_{50}$). Moreover, the distribution of rainfall rates can be considered when computing the probability of rainfall exceeding some predefined threshold, such as the rain rate associated with a 50-yr annual recurrence interval (or return period). Similarly, the probability of exceeding flash flood guidance values can be readily computed. Last, the GAMLSS-produced error model provides the basis to generate ensemble QPEs by sampling the rainfall rate probability distribution functions at each grid point. Techniques make use of LU decomposition to generate equally probable rainfall fields that have spatially and temporally correlated residual errors (Germann et al. 2009; Villarini et al. 2009). The production of ensemble QPEs can lend themselves quite useful to forecast systems such as distributed hydrologic models that forecast the probability of flash flooding.

Problem Sets

Q1: What is probabilistic QPE (PQPE)? Why do we need uncertainty estimation for radar QPEs?

Q2: The radar measures an effective reflectivity factor of 30 dBZ at 50 km. (a) Calculate the rain rate by using the Marshall–Palmer Z–R relation. (b) The radar "sees" 30 dBZ at a range of 100 km. But, because of beam blockage only half of the resolution volume is illuminated. What would have been the correct reflectivity factor without beam blockage? How would your result change depending on which portion of the beam was blocked? Compute the impact of this beam blockage on estimated rainfall rates.

References

Andrieu, H., and J. D. Creutin. 1995. Identification of vertical profiles of radar reflectivity for hydrological applications using an inverse method. Part I: Formulation. *Journal of Applied Meteorology* 34 (1): 225–239.

Andrieu, H., G. Delrieu, and J. Creutin. 1995. Identification of vertical profiles of radar reflectivity for hydrological applications using an inverse method. Part II: Sensitivity analysis and case study. *Journal of Applied Meteorology* 34: 240–259.

Atlas, D. 2002. Radar calibration: Some simple approaches. *Bulletin of the American Meteorological Society* 83 (9): 1313–1316.

Bolen, S. M., and V. Chandrasekar. 2000. Quantitative cross validation of space-based and ground-based radar observations. *Journal of Applied Meteorology* 39 (12): 2071–2079.

Borga, M., E. N. Anagnostou, and E. Frank. 2000. On the use of real-time radar rainfall estimates for flood prediction in mountainous basins. *Journal of Geophysical Research: Atmospheres* (1984–2012) 105 (D2): 2269–2280.

Cho, Y.-H., G. Lee, K.-E. Kim, and I. Zawadzki. 2006. Identification and removal of ground echoes and anomalous propagation using the characteristics of radar echoes. *Journal of Atmospheric and Oceanic Technology* 23 (9): 1206–1222.

Ciach, G. J. 2003. Local random errors in tipping-bucket rain gauge measurements. *Journal of Atmospheric and Oceanic Technology* 20 (5): 752–759.

Ciach, G. J., W. F. Krajewski, and G. Villarini. 2007. Product-error-driven uncertainty model for probabilistic quantitative precipitation estimation with NEXRAD data. *Journal of Hydrometeorology* 8 (6): 1325–1347.

Droegemeier, K., J. Smith, S. Businger et al. 2000. Hydrological aspects of weather prediction and flood warnings: Report of the Ninth Prospectus Development Team of the US Weather Research Program. *Bulletin of the American Meteorological Society* 81 (11): 2665–2680.

Germann, U., and J. Joss. 2002. Mesobeta profiles to extrapolate radar precipitation measurements above the Alps to the ground level. *Journal of Applied Meteorology* 41 (5): 542–557.

Germann, U., M. Berenguer, D., Sempere-Torres, and M. Zappa. 2009. REAL-ensemble radar precipitation estimation for hydrology in a mountainous region. *Quarterly Journal of the Royal Meteorological Society* 135 (639): 445–456.

Giangrande, S. E., J. M. Krause, and A. V. Ryzhkov. 2008. Automatic designation of the melting layer with a polarimetric prototype of the WSR-88D radar. *Journal of Applied Meteorology and Climatology* 47 (5): 1354–1364.

Gourley, J. J., and C. M. Calvert. 2003. Automated detection of the bright band using WSR-88D data. *Weather and Forecasting* 18 (4): 585–599.

Gourley, J. J., B. Kaney, and R. A. Maddox, 2003: Evaluating the calibrations of radars: A software approach. Preprint *Thirty-First International Conference on Radar Meteorology.* Seattle, WA, Amer. Meteor. Soc., 459–462.

Gourley, J. J., D. P. Jorgensen, S. Y. Matrosov, and Z. L. Flamig. 2009. Evaluation of incremental improvements to quantitative precipitation estimates in complex terrain. *Journal of Hydrometeorology* 10 (6): 1507–1520.

Gourley, J. J., P. Tabary, and J. Parent du Chatelet. 2007. A fuzzy logic algorithm for the separation of precipitating from nonprecipitating echoes using polarimetric radar observations. *Journal of Atmospheric and Oceanic Technology* 24 (8): 1439–1451.

Habib, E., A. V. Aduvala, and E. A. Meselhe. 2008. Analysis of radar-rainfall error characteristics and implications for streamflow simulation uncertainty. *Hydrological Sciences Journal* 53 (3): 568–587.

Joss, J., J. Thams, and A. Waldvogel. 1968. The accuracy of daily rainfall measurements by radar. *Proceedings of the 13th Radar Meteorology Conference.* Montreal, Amer. Meteor. Soc., 448–451.

Kirstetter, P.-E., G. Delrieu, B. Boudevillain, and C. Obled. 2010. Toward an error model for radar quantitative precipitation estimation in the Cévennes–Vivarais region, France. *Journal of Hydrology* 394 (1): 28–41.

Kirstetter, P.-E., Y. Hong, J. Gourley et al. 2012. Toward a framework for systematic error modeling of spaceborne precipitation radar with NOAA/NSSL ground radar-based national mosaic QPE. *Journal of Hydrometeorology* 13 (4): 1285–1300.

Kirstetter, P. E., H. Andrieu, B. Boudevillain, and G. Delrieu. 2013a. A physically based identification of vertical profiles of reflectivity from volume scan radar data. *Journal of Applied Meteorology and Climatology* 52 (7): 1645–1663.

Kirstetter, P. E., N. Viltard, and M. Gosset. 2013b. An error model for instantaneous satellite rainfall estimates: Evaluation of BRAIN-TMI over West Africa. *Quarterly Journal of the Royal Meteorological Society* 139 (673): 894–911.

Kirstetter, P.-E., Y. Hong, J. J. Gourley, M. Schwaller, W. Petersen, and Q. Cao. 2014. Impact of sub-pixel rainfall variability on spaceborne precipitation estimation: Evaluating the TRMM 2A25 product, *Quarterly Journal of the Royal Meteorological Society* (Accepted).

Kitchen, M., R. Brown, and A. Davies. 1994. Real-time correction of weather radar data for the effects of bright band, range and orographic growth in widespread precipitation. *Quarterly Journal of the Royal Meteorological Society* 120 (519): 1231–1254.

Legates, D. R., and T. L. DeLiberty. 1993. Precipitation measurement biases in the United States. *Journal of the American Water Resources Association* 29 (5): 855–861.

Liu, H., and V. Chandrasekar. 2000. Classification of hydrometeors based on polarimetric radar measurements: Development of fuzzy logic and neuro-fuzzy systems, and in situ verification. *Journal of Atmospheric and Oceanic Technology* 17 (2): 140–164.

Maddox, R. A., J. Zhang, J. J. Gourley, and K. W. Howard. 2002. Weather radar coverage over the contiguous United States. *Weather and Forecasting* 17 (4): 927–934.

Mandapaka, P. V., and U. Germann. 2010. Radar-rainfall error models and ensemble generators. *Geophysical Monograph Series* 191: 247–264.

Marsalek, J. 1981. Calibration of the tipping-bucket raingage. *Journal of Hydrology* 53 (3): 343–354.

Marshall, J. S., and W. M. K. Palmer. 1948. The distribution of raindrops with size. *Journal of Meteorology* 5 (4): 165–166.

Matrosov, S. Y., K. A. Clark, and D. E. Kingsmill. 2007. A polarimetric radar approach to identify rain, melting-layer, and snow regions for applying corrections to vertical profiles of reflectivity. *Journal of Applied Meteorology and Climatology* 46 (2): 154–166.

Minder, J. R., D. R. Durran, and G. H. Roe. 2011. Mesoscale controls on the mountainside snow line. *Journal of the Atmospheric Sciences* 68 (9): 2107–2127.

Nicol, J., P. Tabary, J. Sugier, J. Parent-du-Chatelet, and G. Delrieu. 2003. Non-weather echo identification for conventional operational radar. Preprints, *Proceedings of 31st International Conference on Radar Meteorology.* Seattle, WA, Amer. Meteor. Soc., 542–545.

Nystuen, J. A. 1999. Relative performances of automatic rain gauges under different rainfall conditions. *Journal of Atmospheric and Oceanic Technology* 16 (8): 1025–1043.

Rigby, R. A., and D. M. Stasinopoulos. 2005. Generalized additive models for location, scale and shape. *Journal of the Royal Statistical Society: Series C (Applied Statistics)* 54 (3): 507–554.

Seo, D.-J., J. Breidenbacii, R. Fulton, and D. Miller. 2000. Real-time adjustment of range-dependent biases in WSR-88D rainfall estimates due to nonuniform vertical profile of reflectivity. *Journal of Hydrometeorology* 1 (3): 222–240.

Silverman, B. W. 1986. *Density Estimation for Statistics and Data Analysis.* Vol. 26. Boca Raton, FL: CRC Press.

Smith, C. J. 1986. The reduction of errors caused by bright bands in quantitative rainfall measurements made using radar. *Journal of Atmospheric and Oceanic Technology* 3 (1): 129–141.

Tabary, P. 2007. The new French operational radar rainfall product. Part I: Methodology. *Weather and Forecasting* 22 (3): 393–408.

Vignal, B., H. Andrieu, and J. D. Creutin. 1999. Identification of vertical profiles of reflectivity from volume scan radar data. *Journal of Applied Meteorology* 38 (8): 1214–1228.

Villarini, G., and W. F. Krajewski. 2010. Review of the different sources of uncertainty in single polarization radar-based estimates of rainfall. *Surveys in Geophysics* 31 (1): 107–129.

Villarini, G., W. F. Krajewski, G. J. Ciach, and D. L. Zimmerman. 2009. Product-error-driven generator of probable rainfall conditioned on WSR-88D precipitation estimates. *Water Resources Research* 45 (1): W01404.

Wilson, J. W., and E. A. Brandes. 1979. Radar measurement of rainfall: A summary. *Bulletin of the American Meteorological Society* 60 (9): 1048–1058.

Zawadzki, I. 1975. On radar-raingauge comparison. *Journal of Applied Meteorology* 14 (8): 1430–1436.

Zhang, J., K. Howard, and J. Gourley. 2005. Constructing three-dimensional v-radar reflectivity mosaics: Examples of convective storms and stratiform rain echoes. *Journal of Atmospheric and Oceanic Technology* 22 (1): 30–42.

Zrnić, D. A. S., A. Ryzhkov, J. Straka, Y. Liu, and J. Vivekanandan. 2001. Testing a procedure for automatic classification of hydrometeor types. *Journal of Atmospheric and Oceanic Technology* 18 (6): 892–913.

3

Polarimetric Radar Quantitative Precipitation Estimation

Thus far we have introduced basic radar principles and processing steps for computing quantitative precipitation estimates (QPE) using conventional, single-polarization radar. The pulse is typically polarized about the horizontal plane, and the primary measurement used for QPE is radar reflectivity, Z. In addition to challenges with data quality (i.e., contamination by nonweather scatterers), many studies have shown that Z alone is insufficient to reveal the natural variability of precipitation (Battan 1973; Rosenfeld and Ulbrich 2003). The drop size distribution (DSD) exhibits variability and thus cannot be adequately described using a single reflectivity-to-rainfall rate (Z–R) relation. With the development of dual-polarization radar (also called polarimetric radar), the accuracy of QPE has been improved through the use of polarimetric varaibles (Bringi and Chandrasekar 2001). The U.S. Next-Generation Radar (NEXRAD) network has been upgraded with dual-polarization technology, and similar upgrades have been conducted or are planned in many other countries. This chapter presents the QPE approaches using polarimetric radar measurements. Issues of radar data quality control and hydrometeor classification, which are critical to obtain accurate radar QPE, are also addressed.

3.1 Polarimetric Radar Variables

This section introduces the essential polarimetric radar variables that are used in QPE. Radar echoes are combined signals backscattered by all the hydrometeors within a radar resolution volume at a given range gate. The intensity and phase of received radar echoes are determined by both scattering and propagation effects. These effects depend on the radar frequency and the size, intensity, phase, shape, structure, and orientation of the hydrometeors. The theoretical equations for the polarimetric radar variables are given below. The use of subscripts for polarimetric variables is quite common. In general, letters in lowercase correspond to linear units, while those in uppercase correspond to units in dB.

1. Radar reflectivity factors at horizontal and vertical polarizations ($Z_{h,v}$ or $Z_{H,V}$)

$$Z_{h,v} = \frac{4\lambda^4}{\pi^4 |K|^2} \int_{D_{min}}^{D_{max}} |f_{h,v}(\pi, D)|^2 N(D) dD \ \left(mm^6 m^{-3}\right) \tag{3.1}$$

$$Z_{H,V} = 10 \log_{10}(Z_{h,v}) \ \ (dBZ) \tag{3.2}$$

2. Differential reflectivity (Z_{dr} or Z_{DR})

$$Z_{dr} = Z_h / Z_v, \tag{3.3}$$

$$Z_{DR} = 10 \log_{10}(Z_h / Z_v) = Z_H - Z_V \ \ (dB) \tag{3.4}$$

3. Co-polar correlation coefficient (ρ_{hv})

$$\rho_{hv} = \frac{\int_{D_{min}}^{D_{max}} f_h^*(\pi, D) f_v(\pi, D) N(D) dD}{\sqrt{\int_{D_{min}}^{D_{max}} |f_h(\pi, D)|^2 N(D) dD \int_{D_{min}}^{D_{max}} |f_v(\pi, D)|^2 N(D) dD}} \tag{3.5}$$

4. Specific differential phase (K_{dp})

$$K_{dp} = \frac{180\lambda}{\pi} \int_{D_{min}}^{D_{max}} \text{Re}\left[f_h(0, D) - f_v(0, D)\right] N(D) dD \ \ (deg \ km^{-1}) \tag{3.6}$$

5. Differential phase (Φ_{dp})

$$\Phi_{dp}(r_g) = 2 \int_0^{r_g} K_{dp}(r) dr \ \ (deg) \tag{3.7}$$

6. Specific attenuation at horizontal or vertical polarization (A_H or A_V)

$$A_{H,V} = 8.686\lambda \int_{D_{min}}^{D_{max}} \text{Im}\left[f_{h,v}(0, D)\right] N(D) dD \ \ (dB \ km^{-1}) \tag{3.8}$$

7. Specific differential attenuation (A_{DP})

$$A_{DP} = A_H - A_V \ \ (dB \ km^{-1}) \tag{3.9}$$

In Equations (3.1)–(3.9), λ is the radar wavelength; $K = (\varepsilon_r - 1)/(\varepsilon_r + 2)$, where ε_r is the complex dielectric constant of water; D denotes the effective diameter of particle (i.e., hydrometeor); D_{max} (or D_{min}) indicates the maximum (or minimum) D within a radar resolution volume; and $N(D)$ is the particle size distribution (PSD) of all these particles; $f_{h,v}$ represents the complex scattering amplitude at horizontal or vertical polarization and the parameters 0 and π for $f_{h,v}$ denote the forward-scattering and backward-scattering components, respectively; the notation $|.|$ signifies the complex norm and Re (or Im) indicates the real (or imaginary) part of a complex number; and r denotes the range from radar and r_g is the range for a given range gate.

Z_h represents the energy backscattered by precipitating hydrometeors and depends on their concentration, size, and phase, which have a close connection to precipitation rate and water content. Z_{dr} is directly related to the median size of observed hydrometeors, a parameter used to describe the DSD, and thus provides valuable supplementary information for QPE. K_{dp} is dependent on the raindrop number concentration but is less sensitive to the size distribution than Z_h. It is independent of radar calibration and partial beam blockage and relatively immune to hail contamination in rain estimation. Positive K_{dp} values result from a phase lag in the horizontally polarized wave compared with the vertical one. Oblate raindrops (those that have larger horizontal dimensions than vertical) basically cause a slight phase delay, which is more pronounced at horizontal polarization. These three polarimetric measurements can be directly applied for estimating rainfall. The correlation coefficient (ρ_{hv}) indicates how well the backscatter amplitudes at vertical and horizontal polarization are correlated. It is a good indicator of hydrometeor phase (homogenous vs. mixed phase) and data quality. This variable is used for classifying the hydrometeor species of the radar echo, which benefits QPE. Precipitation can cause strong attenuation (power loss) in radar measurements, depending on the frequency of the radar wave. Specific attenuation (A_V, A_H) and specific differential attenuation (A_{DP}) are two important variables to address how much power has been lost in Z_h, Z_v, or Z_{dr}, though they are not directly measured. If the attenuation effect is not negligible such as with C-, X-, and Ka/Ku-band radars, attenuated Z_h and Z_{dr} need to be corrected to avoid underestimation in QPE. Values of A_H and A_{DP} also have a strong correlation with precipitation rate.

3.2 Polarimetric Radar Data Quality Control

The quality of radar data is essential to the performance of various radar applications. For example, a 3 dB error in reflectivity may cause 100% overestimation of precipitation. Data quality can be degraded by many factors such as system noise, clutter, attenuation, and miscalibration. Therefore, careful data quality control is required with radar measurements, especially for QPE. This section addresses the recent progress in polarimetric radar data quality control.

3.2.1 Noise Effect and Reduction

System noise has a ubiquitous effect on radar data and is one of most common error sources. The noise can change the quantity of measurements, resulting in different physical interpretation of polarimetric data. For example, the ρ_{hv} of rainfall or dry snow or ice should be close to 1. System noise can reduce the value of ρ_{hv} to below 0.9, which might lead to an incorrect interpretation that raindrops are either mixed-phase precipitation or even ground clutter. Conventional methods for reducing noise are mainly based on the processing of level II radar data, i.e., the moment data. A common approach to mitigate the noise is the smoothing of moment data. For example, the estimation of K_{dp} is usually done by a gradient calculation of averaging Φ_{DP} over multiple range gates, as described in Ryzhkov et al. (2005a). Hubbert and Bringi (1995) applied a low-pass filter to smooth the Φ_{dp} measured along a radar beam path. Other measurements such as Z_h and Z_{dr} are usually smoothed (e.g., over 1 km range) as well before they are used for the QPE. Lee et al. (1997) introduced a speckle filter technique. This method can spatially smooth the radar measurements and improve their utility in applications (Cao et al. 2010). Generally, smoothing data can have the effect of worsening their resolution. However, it helps to obtain a better precipitation estimation with smaller variance.

The noise effect can also be mitigated by advanced radar signal processing, which occurs with the "raw" level I data, i.e., time-series data. The conventional autocorrelation/cross-correlation function (ACF/CCF) method gives the moment estimation mainly based on lag-0 of ACF/CCF. Considering the lag-0 is primarily affected by noise while other lags are not, Melnikov (2006) and Melnikov and Zrnić (2007) have proposed the lag-1 estimator for polarimetric radar variables. To sufficiently apply the information of ACF and CCF, Lei et al. (2012) proposed the multilag correlation estimator for radar moment data estimation. Cao et al. (2012) integrated the multilag processing into the spectrum-time estimation and processing (STEP) algorithm and effectively improved the quality of polarimetric radar data by reducing the noise effect.

3.2.2 Clutter Detection and Removal

Ground clutter generally comes from stationary targets and has a small radial velocity in its radar measurements. To remove it, conventional radar systems usually apply various "notch" filters such as finite/infinite impulse response (FIR/IIR) filters to detect echoes with 0 Doppler velocity (e.g., Torres and Zrnić 1999). A clutter filter is designed to process the time-series data and is generally easy to implement in a radar system. However, the filter can erroneously remove weather signals as well if the weather component also has a small radial velocity. Advanced clutter filtering techniques are mostly based on spectrum analysis. The most popular one is the Gaussian model adaptive processing (GMAP) algorithm introduced by Siggia and Passarelli (2004). It can reconstruct the weather component, which might be removed by a notch filter. Cao et al. (2012) have proposed another spectrum-based algorithm, STEP, which models weather and clutter components in the spectrum and estimates them with a regression approach. Nguyen et al. (2008) have introduced a parametric time domain method (PTDM), which models the weather and clutter component in the ACF. These advanced algorithms are superior to clutter filters because they can preserve the weather components while removing ground clutter, especially when clutter and weather components have similar radial velocities.

These advanced clutter-filtering methods normally require iterative computations and may not be applied everywhere. Therefore, clutter identification is highly desirable for efficient filtering. A typical algorithm is the clutter mitigation decision (CMD) algorithm developed by the National Center for Atmospheric Research (NCAR), which is mainly based on the phase of clutter signal (Hubbert et al. 2009). The spatial continuity of weather signals in the range-spectrum space has also been applied to identify clutter (Morse et al. 2002). Recently Moisseev and Chandrasekar (2009) proposed a novel spectral algorithm, which applied the dual-polarization spectral decomposition to identify clutter. Li et al. (2012) have also introduced a different spectrum clutter identification (SCI) algorithm, which examines both the power and phase information in the spectral domain. The detection and removal of ground clutter are of great significance to the application of polarimetric radar data.

3.2.3 Attenuation Correction

Precipitation attenuation is an unavoidable problem in radar QPE. Fortunately, power losses due to precipitation attenuation at S band, which is the frequency used for the U.S. NEXRAD network, is not significant for most situations. However, it is a major problem for shorter wavelength radars (e.g, C, X, Ku, and Ka band), and thus requires correction. Figure 3.1 shows X-band dual-polarimetric variables of Z_H, Z_{DR}, Φ_{dp}, and ρ_{hv} in a heavy convective cell in the south of France. At this frequency, we can see that values of Φ_{dp} exceeded 160 deg. In the same region, we see Z_H and Z_{DR} became biased very low. The negative values of Z_{DR} are strong indicators that the data are biased because Z_{DR} should be nonnegative in liquid rain due to the spherical shapes of small droplets and oblate shapes with larger drops. The strong gradients shown in Figure 3.1 are due to attenuation loss rather than real spatial

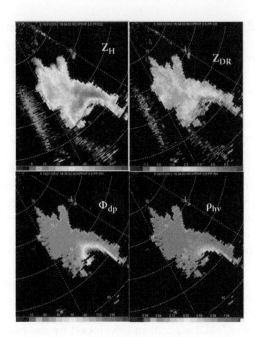

FIGURE 3.1
Polarimetric variables at X-band for a heavy convective cell observed in the south of France on October 21, 2012, at 1834 UTC. Attenuation is noticeable with losses in Z_H, Z_{DR}, and ρ_{hv}, and an associated increase in Φ_{dp}.

gradients in the storm structure. The signal was lost in this storm due to too much attenuation of the signal by the raindrops. Up to the range before the signal was completely lost and unrecoverable, the loss in power in Z_H and Z_{DR} is related to the increase in Φ_{dp}.

Previous algorithms to correct for attenuation losses with single-polarization radars are mainly based on the Hitschfeld–Bordan algorithm and its revised version (e.g., Delrieu et al. 2000). These algorithms rely on the empirical power-law relation between attenuation and radar reflectivity. The nonattenuated reflectivity at each range gate is iteratively computed through keeping the consistency with the attenuated observations along the radar beam path. Because K_{dp} has a strong correlation with A_H (or A_{DP}), their relations (generally the power-law relation) are usually applied to estimate the nonattenuated Z_h or Z_{dr}. Simple methods generally assume known exponents and coefficients in these relations (Bringi et al. 1990; Matrosov et al. 2002). Complex methods consider the dependence of these exponents and coefficients on various factors, such as drop temperature, drop shape model, and DSD variation (Bringi et al. 2001; Park et al. 2005; Gorgucci and Baldini 2007). Attenuation can also be corrected with optimal retrieval approaches such as the variational approach (Hogan 2007; Xue et al. 2009; Cao et al. 2013a).

The general form of the equations for the first-order corrections to Z_H and Z_{DR} due to precipitation attenuation is related to the path-integrated differential phase measurements (Φ_{dp}) as

$$\Delta Z_H = a\Phi_{dp} \ \ \text{(dB)} \tag{3.10}$$

$$\Delta Z_{DR} = b\Phi_{dp} \ \ \text{(dB)} \tag{3.11}$$

where the coefficients a and b (in dB deg^{-1}) depend on a number of the afore-mentioned factors. The most important one is the radar frequency. At S band, Ryzhkov and Zrnić (1995) estimated a and b values to be 0.040 dB deg^{-1} and 0.0083 dB deg^{-1}, respectively. Carey et al. (2000) conducted a literature review for a and b values at C band and synthesized mean values of 0.0688 dB deg^{-1} and 0.01785 dB deg^{-1}. The constants increase with shorter radar wavelength. At X band, Matrosov et al. (2002) found values of 0.22 dB deg^{-1} and 0.032 dB deg^{-1} for a and b, respectively. Clearly, caution must be exercised when using values of Z_H and Z_{DR} for rainfall estimation if they have been heavily corrected due to attenuation loss. This problem increases with decreasing radar wavelength and with heavy rain.

3.2.4 Calibration

Compared with single-polarization radar QPE, polarimetric radar QPE is more sensitive to data quality. Z_{dr} has a small dynamic range for hydrometeor measurements. A change of only several tenths of a dB in Z_{DR} may cause significant changes in QPE. Therefore, calibrating the systematic bias is extremely important for Z_{dr}. The basic method is the engineering calibration, which will accurately measure and compare the gain/damping within the receiving paths for two polarimetric channels. This approach is most reliable but not appropriate for a frequent routine. The antenna can be pointed vertically in light precipitation. Since small raindrops have a spherical shape when they are viewed by radar vertically, both H and V polarimetric channels will receive very similar backscattered signals. That is to say, the Z_{DR} should approach zero at vertical incidence. Another method is to track the sun because Z_{DR} should be zero as well for sun signals. The National Severe Storm Laboratory (NSSL) has applied this method for the calibration of the polarimetric NEXRAD network (Zrnić et al. 2006).

3.2.5 Self-Consistency Check

Ryzhkov et al. (2005b) showed that Z_H in rain could be approximated from Z_{DR} and K_{dp} measurements using the following relation:

$$Z_H = a + b \log (K_{dp}) + cZ_{DR} \tag{3.12}$$

TABLE 3.1

Coefficients for a Third-Degree Polynomial
Fit to the Polarimetric Consistency Relations
at Three Weather Radar Frequencies

Frequency	a_0	a_1	a_2	a_3
X-band	11.74	−4.020	−0.140	0.130
C-band	6.746	−2.970	0.711	−0.079
S-band	3.696	−1.963	0.504	−0.051

where the coefficients a, b, and c depend on radar wavelength and prevalent raindrop shape. These coefficients are also supposed to be relatively insensitive to the raindrop size distribution. Similarly, other researchers (e.g., Vivekanandan et al. 2003) showed that K_{dp} could be expressed as a function of Z_H and Z_{DR}. The self-consistency check uses the dependence between Z_H, Z_{DR}, and K_{dp} to assess the calibration of the radar, although it has many limitations compared with other calibration methods previously mentioned. For example, K_{dp} can be quite noisy in light rain and the calibration may be suitable for moderate or heavy rain only. Despite its limitations, the self-consistency check can be applied to any polarimetric radar by observing rain. No additional costs are associated with other instruments.

Gourley et al. (2009) employed the drop shape model of Brandes et al. (2002), assumed a drop temperature of 0°C, used a normalized gamma model for the DSD, and computed third-order polynomial regressions for the consistency relations at X-, C-, and S-band frequencies. Their regression takes the following form:

$$\frac{K_{dp}}{Z_h} = 10^{-5}(a_0 + a_1 Z_{DR} + a_2 Z_{DR}^2 + a_3 Z_{DR}^3) \tag{3.13}$$

where K_{dp} is one way in units of deg km^{-1}, Z_h is in linear units (mm^6 m^{-3}), and Z_{DR} is in dB. Table 3.1 provides the values for the coefficients as a function of the radar frequency. Application of self-consistency theory to radar observations in rain can be quite useful to diagnose miscalibrated radar observations, in particular Z_h.

3.3 Hydrometeor Classification

The scattering characteristics for hydrometeors can differ significantly (Straka et al. 2000; Park et al. 2009). Z_h for solid particles (graupel, snow aggregate, or hail) is much lower than raindrops with the same water content while melting snow/hail has larger Z_h than its liquid phase counterpart that has melted completely. Storms rarely have a single hydrometeor species within them. Typically, raindrops that reach the surface have their origins in melting snowflakes aloft, followed by pristine ice above that. These raindrops can be mixed with graupel, hail, and many other nonweather scatterers (e.g., birds, insects) that contribute to the radar signal. Theoretically, appropriate scattering models should be applied to the different hydrometeor species to yield accurate QPE. Thus, it is important to identify the different hydrometeor species within the radar volume to guide the QPE algorithms.

3.3.1 Polarimetric Characteristics of Radar Echoes

The basis for hydrometeor classification is the different scattering characteristics of various targets measured by polarimetric radar. For those scatterers that are approximately spherical in shape (e.g., small raindrops) or behave like isotropic scatterers (e.g., dry, tumbling hail), Z_{DR} and K_{dp} values are close to zero. Z_{DR} and K_{dp} values increase as the particle sizes increase. Taking S-band polarimetric radar for example, Z_{DR} (K_{dp}) values normally increase from 0 to 5 dB (3 deg km^{-1}) for drizzle, tropical rain, weak convective rain, stratiform rain, and intensive convective rain. The increase of Z_{DR} and K_{dp} values follow increases in the median size and concentration of the raindrops. Table 3.2 gives some typical ranges of polarimetric variables (S-band) for different radar echoes. It suggests that different radar echoes have very distinguishable polarimetric signatures that contribute to their identification.

3.3.2 Classification Algorithms

The fuzzy logic scheme introduced in Section 2.2.2 has great flexibility and accommodates multiple polarimetric radar measurements to identify different hydrometeors and nonhydrometeors. Popular algorithms include the

TABLE 3.2

Typical Ranges of Polarimetric Variables (S-Band) for Different Radar Echoes

Category	Z_H (dBZ)	Z_{DR} (dB)	K_{dp} (degree/km)	ρ_{hv}
Rain (light, moderate, heavy)	5–55	0–5	0–3	0.98–1.0
Graupel	25–50	0–0.5	0–0.2	0.97–0.995
Dry hail	45–75	–1–1	–0.5–0.5	0.85–0.97
Melting hail	45–75	1–7	–0.5–1	0.75–0.95
Ice crystal	<30	<4	–0.5–0.5	0.98–1.0
Dry snow aggregate	<35	0–0.3	0–0.05	0.97–1.0
Wet snow aggregate	<55	0.5–2.5	0–0.5	0.9–0.97
Ground clutter	20–70	–4–2	very noisy	0.5–0.95
Biological scatterer	5–20	0–12	low & very noisy	0.5–0.8

radar echo classifier (REC) developed by NCAR (Vivekanandan et al. 1999), the polarimetric hydrometeor classification algorithm (HCA) developed by NSSL (Straka et al. 2000; Park et al. 2009), and the hydrometeor classification system (HCS) developed by Colorado State University (Lim et al. 2005). These algorithms generally classify more than 10 distinct species of radar echoes, such as rain, snow, hail, clutter, and so on. Their inputs include not only polarimetric measurements but also the texture and/or error information of these measurements. Other information, such as a temperature profile and radial velocity, is used as well in some algorithms. The latest version of HCA is described in detail by Park et al. (2009). The HCA uses six radar variables for classification: (1) Z_H, (2) Z_{DR}, (3) ρ_{hv}, (4) K_{dp}, (5) texture of Z_H, and (6) texture of Φ_{dp}. The HCA discriminates between 10 classes of radar echo: (1) ground clutter including anomalous propagation (GC/AP), (2) biological scatterers (BS), (3) dry aggregated snow (DS), (4) wet snow (WS), (5) crystals of various orientations (CR), (6) graupel (GR), (7) big drops (BD), (8) light and moderate rain (RA), (9) heavy rain (HR), and (10) a mixture of rain and hail (RH).

Gourley et al. (2007) developed membership functions in a fuzzy logic algorithm that were empirically based on radar observations. Generally, this approach does not apply to a diverse array of hydrometeor species because it is much more difficult to isolate a radar data sample to each individual hydrometeor species. The membership functions tend to be less precise as a result and are commonly designed as beta functions or simpler trapezoidal functions, based on the values shown in Table 3.2. The weights for each variable are also manually designated. For example, in the HCA of Park et al. (2009), K_{dp} is only considered for the classification of CR, HR, and RH. For other classes, the polarimetric signature of K_{dp} is not essential and can be ignored. ρ_{hv} is a primary discriminator for GC/AP, BS, and WS. The texture parameters of Z_H and Φ_{dp} fields are also major contributors to the identification of GC/AP and BS. Examination of data plotted in the Z_H–Z_{DR} plane serves as the basis for discriminating most hydrometeors.

Data quality is an essential issue for the radar echo classification, which may be degraded by biased or erroneous measurements. The HCA algorithm of Park et al. (2009), which is the algorithm running on the NEXRAD polarimetric radars, has introduced an additional weighting parameter called a confidence vector in the aggregation function to address the impact of measurement errors. HCA considers several error factors, including radar miscalibration, attenuation, nonuniform beam filling (NBF), partial beam blockage (PBB), ρ_{hv}, and signal-to-noise ratio (SNR), which are either sources or indicators of measurement errors (Bringi and Chandrasekar 2001; Ryzhkov 2007; Giangrande and Ryzhkov 2005). If the quality of a given radar measurement is deemed more erroneous, then this measurement receives lower weight in the classification scheme.

The classification of hydrometeors is greatly enhanced by considering the temperature profile. This information helps guide the algorithm in terms of precipitation phases. Solid and liquid hydrometeors occasionally have similar polarimetric signatures but they exist at much different temperatures. A good example can be seen with the radar measurements of dry snow aggregates and light rain in Table 3.2. In this case, when the temperature profile is considered, the ambiguity of their classification can be well resolved. There are distinct physical processes below, above, and within the melting layer (Fabry and Zawadzki 1995; Cao et al. 2013b). Therefore, the types of hydrometeors in these regions should be physically constrained. It is not reasonable to identify rain above the melting layer or snow well below it. Similarly, biological scatterers are less likely to exist above the melting layer.

Beam broadening and beam center height increasing in altitude with range complicates the HCA functioning. Within a specific range r_b, the radar only measures the rain region (below the melting layer) while beyond a specific range r_t ($r_t > r_b$), the radar only measures solid hydrometeors above the melting layer. Within the range between r_b and r_t, the hydrometeors that the radar measures may come from the rain region, melting layer, and/or ice region. This may increase the ambiguity of the hydrometeor classification within this range. To reduce the classification error, HCA has implemented several rules to confine the radar echo classes as a function of range. Within the range r_b, there are only GC/AP, BS, BD, RA, HR, and RH classes. Beyond the range r_t, HCA only identifies DS, CR, GR, and RH classes. Within the range between r_b and r_t, GC/AP, BS, WS, GR, BD, RA, HR, RH, DS, and CR may be identified. With the physical constraint guided by the height of the melting layer, the ambiguity of classification can be largely reduced.

3.4 Polarimetric Radar-Based QPE

The precipitation rate (R) can be theoretically computed using the following equations:

$$W = \frac{\pi}{6} \times 10^3 \rho \int_{D_{min}}^{D_{max}} D^3 N(D) dD, \quad \left(g\,m^{-3} \right) \tag{3.14}$$

$$R = 6\pi \times 10^{-4} \int_{D_{min}}^{D_{max}} D^3 v(D) N(D) dD, \quad \left(mm\,h^{-1} \right) \tag{3.15}$$

where the units for D, ρ, $v(D)$, and $N(D)$ are mm, g cm^{-3}, m sec^{-1}, and m^{-3} mm^{-1}, respectively. ρ is the density of the hydrometeor (e.g., raindrop). $v(D)$ is its terminal fall velocity, which is mainly dependent on the particle size and density. It also depends on the particle shape and ambient air density. In general, the particle's fall velocity can be approximately represented with a power-law relation as

$$v(D) = aD^b, \quad (\text{m sec}^{-1}) \tag{3.16}$$

Typical values of a vary from 3.6 to 4.2 and b from 0.6 to 0.67. A commonly used power-law relation is $v(D) = 3.78D^{0.67}$ introduced by Atlas and Ulbrich (1977). Atlas et al. (1973) introduced an exponential relation as $v(D) = 9.65 - 10.3\exp(-0.6D)$. Combining different relations and observations in the literature, Brandes et al. (2002) fitted a polynomial relation of a raindrop's falling velocity as $v(D) = -0.1021 + 4.932D - 0.9551D^2 + 0.07934D^3 - 0.002362D^4$. The fall velocities of hailstones and snow aggregates have a large dependence on the particle density and drag force. However, the density of hailstones and snow aggregates may vary for different storms, and their irregular shapes may cause different drag forces as well. Therefore, their fall velocities generally have greater variability than with raindrops. The common relations are $v(D) = 3.62D^{0.5}$ for hailstones (Matson and Huggins 1980) and $v(D) = 0.98D^{0.31}$ for snow aggregates (Gunn and Marshall 1958).

According to Equations (3.14) and (3.15), liquid water content W and precipitation rate R are different moments of the DSD. For example, R is approximately the 3.67th moment of the DSD. Furthermore, Z_H, K_{dp}, and A_H can be expressed by PSD moments as well. Z_H can be approximated by the 6th moment of the DSD under the condition of Rayleigh scattering. K_{dp} in rainfall is approximately proportional to the 4.24th moment for a radar wavelength of 10 cm (Sachidananda and Zrnić 1986). Therefore, power-law relations can usually be found between radar variables (Z_H, K_{dp}, or A_H) and bulk variables (W or R), providing an empirical approach to precipitation estimation.

Many Z_h–R relations have been reported for different rain types, seasons, and locations. Rosenfeld and Ulbrich (2003) gave a complete review of Z_h–R relations and summarized the microphysical processes, which might lead to the Z_h–R variability. Even though radar algorithms have been developed to identify certain characteristic signatures of different rain types, such as convective vs. stratiform echoes, conventional radars cannot directly represent the natural variability of DSDs. Polarimetric variables can be used to observe DSD variability and subsequently improve the accuracy of QPE.

In general, rainfall estimators based on polarimetric radar variables have the following forms:

$$R(Z_h, Z_{dr}) = aZ_h^b Z_{dr}^c \tag{3.17}$$

$$R(K_{dp}) = aK_{dp}^b \tag{3.18}$$

$$R(K_{dp}, Z_{dr}) = aK_{dp}^b Z_{dr}^c \tag{3.19}$$

where R is in mm h^{-1}. Z_h and K_{dp} have linear units in mm^6 m^{-3} and deg km^{-1}, respectively. Z_{dr} is a dimensionless linear ratio.

Table 3.3 gives the parameters *a, b,* and *c* for several common polarimetric radar rainfall estimators at S-, C-, and X-bands. These parameters may change due to different raindrop model assumptions and/or different local DSD climatologies (Bringi et al. 2011; Matrosov 2010). For example, Ryzhkov et al. (2005a) showed the difference of several S-band polarimetric rainfall estimators based on different drop shape models (Pruppacher and Beard 1970; Chuang and Beard 1990; Brandes et al. 2002; Bringi et al. 2003). Despite the differences in the raindrop shape model that was assumed, their study showed that the polarimetric rainfall estimators are less susceptible to DSD variability and generally improved rainfall estimates over the traditional single-polarization relations.

Each polarimetric estimator has its own advantages and disadvantages. The use of Z_{dr} gives a better estimation of raindrops representing the median of the DSD; i.e., those that contribute the majority to the total rainfall amount. However, the dynamic range of Z_{dr}, which is attributed to microphysical variability, is relatively small. It is more susceptible to measurement error and miscalibration than other polarimetric radar variables. Z_{dr} is a relative measurement and must be combined with either Z_h and/or K_{dp} for rainfall estimation. In general, the measurement error of Z_{DR} is on the order of a few tenths of a dB. K_{dp} is a phase measurement and is immune to any error in the absolute calibration of the radar. It is unaffected by precipitation attenuation along the propagation path and less affected by mixed phase precipitation such as rain mixed with hail. However, since K_{dp} is derived from Φ_{dp} measurements over a given path length, K_{dp} estimation error increases rapidly as the path length decreases below 2 km (Bringi and Chandrasekar 2001). This results in a tradeoff between the accuracy and range resolution of K_{dp}. In general, K_{dp} can be estimated to an accuracy of around 0.3–0.4 deg km^{-1} and has a smaller estimation error for heavy rain than for light rain. Therefore, when rainfall is intense and/or mixed with hail, $R(K_{dp})$ is more suitable than other estimators, while in light rain it is not appropriate to apply $R(K_{DP})$ relations.

TABLE 3.3

Parameters for Several Common Polarimetric Radar Rainfall Estimators

	a	b	c	Notes
$R(Z_h, Z_{dr})$	6.7×10^{-3}	0.927	−3.43	S-band (10 cm)
	5.8×10^{-3}	0.91	−2.09	C-band (5.5 cm)
	3.9×10^{-3}	1.07	−5.97	X-band (3 cm)
$R(K_{dp})$	50.7	0.85		S-band (10 cm)
	24.68	0.81		C-band (5.5 cm)
	17.0	0.73		X-band (3 cm)
$R(K_{dp}, Z_{dr})$	90.8	0.93	−1.69	S-band (10 cm)
	37.9	0.89	−0.72	C-band (5.5 cm)
	28.6	0.95	−1.37	X-band (3 cm)

The measurement error of Z_h, Z_{dr}, and K_{dp} may propagate into the final rainfall estimate. The physical variation in the linkage between the polarimetric variables and rainfall rates estimates, which cannot be represented by the estimator error alone, may also result in uncertainty in rainfall estimation. Consequently, the total estimation error is attributed to two terms: ε_m is the error propagating from the measurement and ε_p is the parametric error of the estimator (Bringi and Chandrasekar 2001).

$$\hat{R} = R + \varepsilon_m + \varepsilon_p \tag{3.20}$$

$$\sigma_R^2 = \sigma_m^2 + \sigma_p^2 \tag{3.21}$$

where R denotes the true rainfall rate; notation $\wedge$ indicates the estimation; σ^2 is the error variance; and subscripts R, m, and p indicate the variances associated with the estimation error, measurement error, and parametric error, respectively.

Bringi and Chandrasekar (2001) showed some results of error quantification for different estimators. As for the single polarization estimator $R(Z_h)$, a 0.8 dB measurement error in Z_H results in about 15% uncertainty in R estimation. However, its parametric error introduces a 40% uncertainty given a rainfall rate of 50 mm h^{-1} and a mean uncertainty of 45% for rainfall varying from 1 mm h^{-1} to 250 mm h^{-1}. As for $R(K_{dp})$, the parametric uncertainty is reduced to 25% for rainfall rates of 50 mm h^{-1} and the mean uncertainty is reduced to 27%. For $R(K_{dp},Z_{dr})$, the mean parametric uncertainty is further reduced to about 15%. The same parametric uncertainty exists for $R(Z_h,Z_{dr})$. A 0.8 dB measurement error in Z_H and a 0.2 dB measurement error in Z_{DR} may lead to 24% uncertainty in rainfall estimation. This implies that the use of Z_{dr} adds more measurement error, introducing 9% more uncertainty than $R(Z_h)$. However, $R(Z_h,Z_{dr})$ greatly reduces the estimation uncertainty attributed with the parametric error, which decreases from 45% to 15%.

From the error analysis of different estimators, it can be concluded that the use of polarimetric measurements may enlarge the measurement error effect in the rainfall rate estimators but can effectively reduce the parametric error effect. The overall improvement depends on both factors. Since each estimator has its own limitations, it is desirable to find an optimal way to combine them for better estimation. Ryzhkov et al. (2005a) proposed a "synthetic" estimator for S-band polarimetric radar, which combines various estimators as follows:

$$\begin{cases} R = \overline{R(Z_h)}/f_1(\overline{Z_{dr}}) & \text{if} & \overline{R(Z_h)} < 6 \text{ mm} \cdot \text{h}^{-1} \\ R = \overline{R(K_{dp})}/f_2(\overline{Z_{dr}}) & \text{if} & 6 < \overline{R(Z_h)} < 50 \text{ mm} \cdot \text{h}^{-1} \\ R = \overline{R(K_{dp})} & \text{if} & \overline{R(Z_h)} > 50 \text{ mm} \cdot \text{h}^{-1} \end{cases} \tag{3.22}$$

and

$$\begin{cases} R(Z_h) = 1.7 \times 10^{-2} Z_h^{0.714} \\ R(K_{dp}) = 44.0 \left| K_{dp} \right|^{0.822} \text{sign}\left(K_{dp} \right) \\ f_1(Z_{dr}) = 0.4 + 5.0 \left| Z_{dr} - 1 \right|^{1.3} \\ f_2(Z_{dr}) = 0.4 + 3.5 \left| Z_{dr} - 1 \right|^{1.7} \end{cases} \tag{3.23}$$

where, $||$ means the absolute value and "sign" means the signum function. The mean values $\overline{R(Z_h)}$, $\overline{R(K_{dp})}$, and $\overline{Z_{dr}}$, are computed by averaging over areas of 1 km $\times$ 1 deg.

Ryzhkov et al. (2005a) evaluated different estimators using real radar data and surface rain gauge measurements from April 2002 through July 2003 in Oklahoma. Their study indicates that the polarimetric radar rain estimators have superiority to the single-polarization estimator $R(Z_h)$. There is a large improvement when K_{dp} is used in lieu of Z_h for the rainfall estimation. Additional improvement is seen when Z_{dr} is included. The synthetic estimator $R(Z_h, Z_{dr}, K_{dp})$ shows the best performance because of the optimal use of Z_h, Z_{dr}, and K_{dp}.

3.5 Microphysical Retrievals

Polarimetric radar variables provide more than just rainfall rates. Many other quantities related to hydrometeors such as the characteristic size, particle concentration, and water content are useful for conducting microphysical research, as well as for assimilation in numerical weather prediction models. The particle size distribution (PSD) provides fundamental information on precipitation microphysics, which can be used to calculate the related radar and precipitation variables of interest. This section focuses on rain microphysics. The raindrop size distribution model and retrieval methods are outlined in the following subsections.

3.5.1 Raindrop Size Distribution Model

The DSD retrieval relies on the use of the DSD model, which assumes that natural rain microphysics can be approximately represented by a mathematical function. Researchers in the meteorology community commonly use the following DSD models:

M-P model: $$N(D) = 8000 \exp(-\Lambda D) \tag{3.24}$$

Exponential model: $$N(D) = N_0 \exp(-\Lambda D) \tag{3.25}$$

Gamma model: $N(D) = N_0 D^\mu \exp(-\Lambda D)$ (3.26)

Lognormal model: $N(D) = \dfrac{N_T}{\sqrt{2\pi}\sigma_r D} \exp\left(\dfrac{-[\ln(D) - \eta]^2}{2\sigma_r^2}\right)$ (3.27)

The well-known M-P model (Marshall and Palmer 1948) has been widely used in the past 50 years. It is a single-parameter model with a slope parameter Λ and is helpful in bulk schemes for rain parameterization and rainfall estimation based on single-polarization weather radar. The exponential model is a two-parameter model with slope Λ and a concentration parameter N_0. It is more flexible than the M-P model since the latter is equivalent to the exponential model with a fixed value for N_0. The exponential model is frequently used to model ice/snow PSD. Currently, the gamma model (Ulbrich 1983), a three-parameter model, is widely recognized as the most accurate model to represent the variability of natural DSDs. In addition to N_0 and Λ, the gamma model introduces a shape parameter μ, which varies for different rain types. Some recent studies have applied the normalized gamma DSD model (e.g., Bringi et al. 2002), which introduces DSD parameters related to the bulk variables. The lognormal model also uses three parameters: total number concentration N_T, mean η, and standard deviation σ_r of a Gaussian distribution. This model follows the assumption that the randomness of raindrops can be described as a multivariate Gaussian distribution. It provides a good explanation of DSD based on probability theory. Also, the mathematical calculation is not very complicated. However, it might not be the most suitable model to match observed DSDs.

DSD models mentioned above have their own advantages and limitations. In comparison, the gamma distribution generally has the best performance in modeling observed DSDs. Figure 3.2 shows an example of DSD models. The asterisks denote a DSD observed by a disdrometer. Four lines represent the distributions fitted with four models. It is evident that the M-P model, with only a single parameter, has the least capability of matching the observed DSD. The DSD model with more freedom (i.e., more parameters) represents the observation better. The observed DSD in the figure is well modeled by the gamma function. The cost of this better matching is the requirement to estimate more parameters when retrieving the DSD, as will be shown in the next section.

3.5.2 DSD Retrieval

The rainfall DSD provides fundamental information on rain microphysics. Given a DSD, all the integral parameters describing the properties of rain can be calculated (e.g., R, Z_h, and liquid water content). To retrieve the

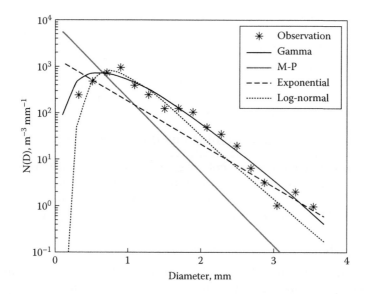

FIGURE 3.2
Example of an observed DSD that is represented by different models.

DSD, one must make an assumption about the DSD model. Although the three-parameter gamma model is more accurate in matching and describing the natural DSD than the other, simpler models, challenges remain in retrieving the parameters. Three-parameter DSD models require independent information from at least three radar measurements. However, the radar measurement error effect with the polarimetric variables might outweigh their contribution. If too much error is introduced in the measurements, then it is better to use fewer variables and a simpler DSD model. In general, Z_h and Z_{dr} are considered as the two most reliable polarimetric measurements for DSD retrievals. Therefore, methods that retrieve two parameters to describe the DSD are typically preferred.

The exponential model is an obvious choice for a two-parameter DSD model. The shape parameter of the DSD, however, has been forced as a constant so that it always assumes that smaller raindrops have greater number concentrations than larger raindrops. This assumption does not apply to all types of rain during certain stages of storm evolution. Some surface observations from disdrometers have shown that very small raindrops (<0.6 mm) have smaller concentrations than raindrops 0.8–1.0 mm in diameter (Cao and Zhang 2009). A gamma model with a fixed and nonzero value for µ can allow the DSD shape to be convex. However, the choice of the µ value is subjective and might not represent the truth. Some studies have shown that the gamma model not only best represents natural DSDs, but that the three gamma parameters (N_0, µ, and Λ) are not mutually independent (Ulbrich 1983; Chandrasekar and Bringi 1987; Haddad et al. 1997; Zhang et al. 2001;

Brandes et al. 2004; Cao et al. 2008). Zhang et al. (2001) found that μ is highly related to Λ and further proposed a constrained-gamma (C-G) DSD model. The C-G model applies an empirical μ-Λ relation, which reduces the gamma model to a two-parameter model while maintaining the flexibility to represent the natural DSDs with the convex shape. Cao et al. (2008) refined the C-G model using disdrometer observations in central Oklahoma. The refined μ-Λ relation is given as

$$\mu = -0.0201\Lambda^2 + 0.902\Lambda - 1.718 \tag{3.28}$$

Given two radar measurements (e.g., Z_h and Z_{dr}) and a two-parameter DSD model, it is straightforward to retrieve DSD parameters according to Equations (3.1)–(3.4), given there are two unknowns and two measurements. After the DSD is retrieved, R can be computed following Equation (3.14).

3.5.3 Snowfall and Hail Estimation

The estimation of snowfall and hail is more difficult than rainfall estimation. This is because snow aggregates and ice crystals have quite different particle shapes from the oblate spheroids associated with raindrops. The snowfall rate estimator can be expressed as a power-law relation as

$$Z_e = \alpha R_s^\beta \tag{3.29}$$

where Z_e is the equivalent radar reflectivity factor (in $mm^6\ m^{-3}$) of water drops and R_s is the snowfall rate expressed as the liquid equivalent per unit time (in mm h^{-1}). A widely used snowfall estimator is the $Z_e = 1780R_s^{2.21}$ suggested by Sekhon and Srivastava (1970). Fujiyoshi et al. (1990) compared α and β parameters proposed in different studies and showed that α (or β) varies within the range of 100–3000 (or 1–2.3). Polarimetric radar measurements are generally useful in the identification of snowfall but seldom used for the quantitative estimation of snowfall rate. The major reason is that the complexity of natural snowflakes makes it difficult to accurately model the scattering properties of snowflakes so that Z_{dr} and other polarimetric variables cannot be easily applied.

Polarimetric radar measurements can be used to distinguish hail from rain (Aydin et al. 1986; Depue et al. 2007). The hail differential reflectivity (H_{DR}) is defined as

$$H_{DR} = Z_H - f(Z_{DR}),$$

where

$$f(Z_{DR}) = \begin{cases} 27 & Z_{DR} \leq 0\ \text{dB} \\ 19Z_{DR} + 27 & 0 \leq Z_{DR} \leq 1.74\ \text{dB} \\ 60 & Z_{DR} > 1.74\ \text{dB} \end{cases} \tag{3.30}$$

The above-referenced studies showed that the H_{DR} thresholds of 21 dB and 30 dB were reasonably successful in respectively identifying the regions where large and structurally damaging hailstones were reported. The vertically integrated liquid water content (VIL) is also a good indicator of hail existence and hail size (Amburn and Wolf 1997). A substantial increase of severe hail (size >19 mm) is usually associated with VIL densities greater than 3.5 g m^{-3}. It is noted that VIL can be derived from the empirical power-law relation between radar Z_h and liquid water content.

There is not a widely applicable Z_h–R relation for hail estimation. Torlaschi et al. (1984) derived a relation between equivalent rainfall rate of hail R_H (mm h^{-1}) and the PSD parameter Λ (mm^{-1}), which is given by

$$\Lambda = \ln(88/R_H)/3.45 \tag{3.31}$$

Cheng and English (1983) proposed a relation $N_0 = 115\Lambda^{3.63}$ to model an exponential PSD of hailstones. According to the Rayleigh approximation and Equation (3.31), the empirical relation between radar reflectivity Z (mm^6 m^{-3}) and R_H (mm h^{-1}) is given by

$$Z = 5.38 \times 10^6 \left[\ln(88/R_H)\right]^{-3.37} \tag{3.32}$$

3.5.4 Validation

Radar QPE algorithms are typically validated by in situ observations. Disdrometers are much more useful for the development and evaluation of polarimetric QPE and DSD retrieval algorithms. Disdrometers can be used to validate the DSD assumptions applied in the retrievals. Moreover, they can be used to quantify the DSD variation for various precipitation types (Chang et al. 2009). The conventional disdrometer is the impact type, which is designed based on the measurement of raindrop momentum (Tokay et al. 2001). The most common impact disdrometer is the Joss–Waldvogel (JW) disdrometer. Noted weaknesses of the JW disdrometer include insensitivity to small drops, relatively coarse resolution, and limited measureable size range. More recent disdrometers show enhanced performance by applying optical techniques. Those disdrometers include the one-dimensional laser optical disdrometer (OTT Parsivel disdrometer, Thiess disdrometer), and the two-dimensional video disdrometer (2DVD) (Kruger and Krajewski 2002). Superior to impact-type disdrometers, optical disdrometers can measure the shape and falling velocity of particles. Furthermore, they generally provide more accurate measurements of PSD/DSD.

Problem Sets

QUALITATIVE QUESTIONS

1. What is the difference in Rayleigh scattering and Mie scattering? What is the effect of Mie scattering on the radar measurement of precipitation?

2. Describe the advantages and disadvantages of weather radars in measuring precipitation with different frequencies: X-band (3 cm), C-band (5 cm), and S-band (10 cm). (Hint: Consider the difference in radar/antenna size, transmitter power, radar range, Earth curvature effect, radar resolution, Rayleigh/Mie scattering, precipitation attenuation, etc.)

3. What are the additional measurements provided by polarimetric weather radar, compared with the single-polarization weather radar? What measurements can be applied for QPE, quality control, and radar echo classification?

4. Why is the understanding of hydrometeor shape important for interpreting the polarimetric radar measurement?

5. Why is the particle/raindrop size distribution (PSD/DSD) fundamental for understanding the property of precipitation?

6. Describe the necessity of removing the clutter contamination and correcting the attenuation for radar data. What are their effects on the radar QPE?

7. Why is the particle/raindrop size distribution (PSD/DSD) fundamental for understanding the property of precipitation? What relations are among radar reflectivity, rainfall rate, water content, specific attenuation, and specific differential phase?

8. Compared with single-polarization radar, explain why polarimetric radar measurements help improve the QPE? What are the strengths for different estimators in Section 3.4? (Hint: Discuss the measurement error and model uncertainty.)

QUANTITATIVE QUESTIONS

1. Given the Rayleigh scattering assumption, the radar reflectivity can be approximated by the 6th moment of DSD. There are three measurements of rainfall by S-band radar, with radar reflectivity 25 dBZ, 35 dBZ, and 45 dBZ, and differential reflectivity 0.2 dB, 0.8 dB, and 1.8 dB, respectively.

   a. Calculate the rainfall rate based on the polarimetric estimator $R(Z_h, Z_{dr})$.

   b. Suppose the differential reflectivity values are 0.4 dB, 1.2 dB, and 2.4 dB, respectively. Calculate the rainfall rate again. Explain the difference from the first result.

c. Now assume the M-P model is applied. Do the DSD retrieval, and calculate the rainfall rate based on the retrieved DSD. Compare the results with the previous two results and explain the difference.

2. Suppose the measurement error of K_{dp} is 10%. What is the uncertainty of R estimation in percentage for estimator $R(K_{dp})$? Suppose the measurement errors for Z_h and Z_{dr} are 1 dB and 0.3 dB, respectively. What is the uncertainty of R estimation for estimator $R(Z_h, Z_{dr})$? (Hint: Use approximation $\sigma\left[\hat{X}(dB)\right] \approx 10\log_{10}\left[1+\sigma(\hat{X})/X\right]$.)

3. Zhang et al. (2001) derived empirical relations to quantify the backscattering amplitude of raindrops: $|f_h(\pi, D)| = 4.26 \times 10^{-4} D^{3.02}$ and $|f_v(\pi, D)| = 4.76 \times 10^{-4} D^{2.69}$. There are several DSDs given as follows:

a. Several exponential DSDs with Λ varying with 1, 2, 4, 6, and 8.

b. Several gamma DSDs with Λ of 4 and μ varying with 0.5, 1, 1.5, 2, and 3.

    Calculate Z_{DR} values for these DSDs and describe the change of median size D_0 and/or μ with the Z_{DR}. (Hint: Raindrops are generally less than 8 mm.)

4. Given radar reflectivity and differential reflectivity are 30.1 dBZ and 1.7 dB, respectively, estimate the specific differential phase (degree km^{-1}) for X-band, C-band, and S-band radar measurements. Explain the difference for these radars.

References

Amburn, S. A., and P. L. Wolf. 1997. VIL density as a hail indicator. *Weather and Forecasting* 12: 473–478.

Atlas, D., and C. W. Ulbrich. 1977. Path- and area-integrated rainfall measurement by microwave attenuation in the 1–3 cm band. *Journal of Applied Meteorology* 16: 1322–1331.

Atlas, D., R. C. Srivastava, and R. S. Sekhon. 1973. Doppler radar characteristics of precipitation at vertical incidence. *Reviews of Geophysics and Space Physics* 11: 1–35.

Aydin, K., T. A. Seliga, and V. Balaji. 1986. Remote sensing of hail with a dual-linear polarization radar. *Journal of Climate and Applied Meteorology* 25: 1475–1484.

Battan, L. J. 1973. *Radar Observation of the Atmosphere*. Chicago: University of Chicago Press.

Brandes, E. A., G. Zhang, and J. Vivekanandan. 2002. Experiments in rainfall estimation with a polarimetric radar in a subtropical environment. *Journal of Applied Meteorology* 41: 674–685.

Brandes, E. A., G. Zhang, and J. Vivekanandan. 2004. Drop size distribution retrieval with polarimetric radar: Model and application. *Journal of Applied Meteorology* 43: 461–475.

Bringi, V. N., M. A. Rico-Ramirez, and M. Thurai. 2011. Rainfall estimation with an operational polarimetric C-band radar in the United Kingdom: Comparison with a gauge network and error analysis. *Journal of Hydrometeorology* 12: 935–954.

Bringi, V. N., V. Chandrasekar, J. Hubbert, et al. 2003. Raindrop size distribution in different climatic regimes from disdrometer and dual-polarized radar analysis. *Journal of Atmospheric Science* 60: 354–365.

Bringi, V., G. Huang, V. Chandrasekar, et al. 2002. A methodology for estimating the parameters of a gamma raindrop size distribution model from polarimetric radar data: Application to a squall-line event from the TRMM/Brazil campaign. *Journal of Atmospheric and Oceanic Technology* 19: 633–645.

Bringi, V., and V. Chandrasekar. 2001. *Polarimetric Doppler Weather Radar: Principles and Applications*. Cambridge University Press.

Bringi, V. N., T. D. Keenan, and V. Chandrasekar. 2001. Correcting C-band radar reflectivity and differential reflectivity data for rain attenuation: A self-consistent method with constraints. *IEEE Transactions on Geoscience and Remote Sensing* 39: 1906–1915.

Bringi, V., V. Chandrasekar, N. Balakrishnan, et al. 1990. An examination of propagation effects in rainfall on radar measurements at microwave frequencies. *Journal of Atmospheric and Oceanic Technology* 7: 829–840.

Cao, Q., G. Zhang, E. Brandes, et al. 2008. Analysis of video disdrometer and polarimetric radar data to characterize rain microphysics in Oklahoma. *Journal of Applied Meteorology and Climatology* 47: 2238–2255.

Cao, Q., and G. Zhang. 2009. Errors in estimating raindrop size distribution parameters employing disdrometer and simulated raindrop spectra. *Journal of Applied Meteorology and Climatology* 48: 406–425.

Cao, Q., G. Zhang, E. Brandes, et al. 2010. Polarimetric radar rain estimation through retrieval of drop size distribution using a Bayesian approach. *Journal of Applied Meteorology and Climatology* 49: 973–990.

Cao, Q., G. Zhang, R. Palmer, et al. 2012. Spectrum-time estimation and processing (STEP) for improving weather radar data quality. *IEEE Geoscience and Remote Sensing Letters* 50: 4670–4683.

Cao, Q., G. Zhang, and M. Xue. 2013a. A variational approach for retrieving raindrop size distribution from polarimetric radar measurements in the presence of attenuation. *Journal of Applied Meteorology and Climatology* 52: 169–185.

Cao, Q., Y. Hong, J. J. Gourley, et al. 2013b. Statistical and physical analysis of vertical structure of precipitation in mountainous West Region of US using 11+ year spaceborne TRMM PR observations. *Journal of Applied Meteorology and Climatology* 52: 408–424.

Carey, L. D., S. A. Rutledge, D. A. Ahijevych, and T. D. Keenan. 2000. Correcting propagation effects in C-band polarimetric radar observations of tropical convection using differential propagation phase. *Journal of Applied Meteorology* 39, 1405–1433.

Chandrasekar, V., and V. N. Bringi. 1987. Simulation of radar reflectivity and surface measurements of rainfall. *Journal of Atmospheric and Oceanic Technology* 4: 464–478.

Chang, W. Y., T. C. Wang, and P. L. Lin. 2009. Characteristics of the raindrop size distribution and drop shape relation in typhoon systems in the western Pacific from the 2D video disdrometer and NCU C-band polarimetric radar. *Journal of Atmospheric and Oceanic Technology* 26: 1973–1993.

Cheng, L., and M. English. 1983. A relationship between hailstone concentration and size. *Journal of Atmospheric Science* 40: 204–213.

Chuang, C., and K. Beard. 1990. A numerical model for the equilibrium shape of electrified raindrops. *Journal of Atmospheric Science* 47: 1374–1389.

Delrieu, G., H. Andrieu, and J. D. Creutin. 2000. Quantification of path-integrated attenuation for X- and C-band weather radar systems operating in mediterranean heavy rainfall. *Journal of Applied Meteorology* 39: 840–850.

Depue, T. K., P. C. Kennedy, and S. A. Rutledge. 2007. Performance of the hail differential reflectivity (HDR) polarimetric radar hail indicator. *Journal of Applied Meteorology and Climatology* 46: 1290–1301.

Fabry, F., and I. Zawadzki. 1995. Long-term radar observations of the melting layer of precipitation and their interpretation. *Journal of Atmospheric Science* 52: 838–851.

Fujiyoshi, Y., T. Endoh, T. Yamada, et al. 1990. Determination of a Z–R relationship for snowfall using a radar and high sensitivity snow gauges. *Journal of Applied Meteorology* 29: 147–152.

Giangrande, S. E. and A. V. Ryzhkov. 2005. Calibration of dual-polarization radar in the presence of partial beam blockage. *Journal of Atmospheric and Oceanic Technology* 22: 1156–1166.

Gorgucci, E., and L. Baldini. 2007. Attenuation and differential attenuation correction of C-band radar observations using a fully self-consistent methodology. *IEEE Geoscience and Remote Sensing Letters* 2: 326–330.

Gourley J. J., P. Tabary, and J. Parent-du-Chatelet. 2007. A fuzzy logic algorithm for the separation of precipitating from non-precipitating echoes using polarimetric radar observations. *Journal of Atmospheric and Ocean Technology* 24: 1439–1451.

Gourley, J. J., A. J. Illingworth, and P. Tabary. 2009. Absolute calibration of radar reflectivity using redundancy of polarization observations and implied constraints on drop shapes. *Journal of Atmospheric and Ocean Technology* 26: 689–703.

Gunn, K. L. S., and J. S. Marshall. 1958. The distribution with size of aggregate snowflakes. *Journal of Meteorology* 15: 452–461.

Haddad, Z. S., D. A. Short, S. L. Durden, et al. 1997. A new parameterization of the rain drop size distribution. *IEEE Transactions on Geoscience and Remote Sensing* 35: 532–539.

Hogan, R. J. 2007. A variational scheme for retrieving rainfall rate and hail reflectivity fraction from polarization radar. *Journal of Applied Meteorology and Climatology* 46: 1544–1564.

Hubbert, J. V., and V. N. Bringi. 1995. An iterative filtering technique for the analysis of coplanar differential phase and dual-frequency radar measurements. *Journal of Atmospheric and Oceanic Technology* 12: 643–648.

Hubbert, J. C., M. Dixon, and S. M. Ellis. 2009. Weather radar ground clutter. Part II: Real-time identification and filtering. *Journal of Atmospheric and Oceanic Technology* 26: 1181–1197.

Kruger, A., and W. F. Krajewski. 2002. Two-dimensional video disdrometer: A description. *Journal of Atmospheric and Oceanic Technology* 19: 602–617.

Lee, J. S., M. R. Grunes, and G. De Grandi. 1997. Polarimetric SAR speckle filtering and its impact on classification. *Proceedings of IEEE International Conference on Geoscience and Remote Sensing Symposium* 2: 1038–1040.

Lei, L., G. Zhang, R. J. Doviak, et al. 2012. Multilag correlation estimators for polarimetric radar measurements in the presence of noise. *Journal of Atmospheric and Oceanic Technology* 29: 772–795.

Li, Y., G. Zhang, R. J. Doviak, et al. 2012. A new approach to detect ground clutter mixed with weather signals. *IEEE Transactions on Geoscience and Remote Sensing* 99: 1–15.

Lim, S., V. Chandrasekar, and V. Bringi. 2005. Hydrometeor classification system using dual-polarization radar measurements: Model improvements and in situ verification. *IEEE Transactions on Geoscience and Remote Sensing* 43: 792–801.

Marshall, J. S. and W. M. Palmer. 1948. The distribution of raindrops with size. *Journal of Meteorology* 5: 165–166.

Matrosov, S. Y., K. A. Clark, B. E. Martner, et al. 2002. X-band polarimetric radar measurements of rainfall. *Journal of Applied Meteorology* 41: 941–952.

Matrosov, S. Y. 2010. Evaluating polarimetric X-band radar rainfall estimators during HMT. *Journal of Atmospheric and Oceanic Technology* 27: 122–134.

Matson, R. J., and A. W. Huggins. 1980. The direct measurement of the sizes, shapes and kinematics of falling hailstones. *Journal of Atmospheric Science* 37: 1107–1125.

Melnikov, V. M. 2006. One-lag estimators for cross-polarization measurements. *Journal of Atmospheric and Oceanic Technology* 23: 915–926.

Mishchenko, M. I., L. D. Travis, and D. W. Mackowski. 1996. T-matrix computations of light scattering by nonspherical particles: A review. *Journal of Quantitative Spectroscopy and Radiative Transfer* 55: 535–575.

Moisseev, D. N., and V. Chandrasekar. 2009. Polarimetric spectral filter for adaptive clutter and noise suppression. *Journal of Atmospheric and Oceanic Technology* 26: 215–228.

Nguyen, C. M., D. N. Moisseev, and V. Chandrasekar. 2008. A parametric time domain method for spectral moment estimation and clutter mitigation for weather radars. *Journal of Atmospheric and Oceanic Technology* 25: 83–92.

Park, H., A. V. Ryzhkov, D. S. Zrnić, et al. 2009. The hydrometeor classification algorithm for the polarimetric WSR-88D: Description and application to an MCS. *Weather and Forecasting* 24: 730–748.

Park, S. G., V. N. Bringi, V. Chandrasekar, et al. 2005. Correction of radar reflectivity and differential reflectivity for rain attenuation at X band. Part I: Theoretical and empirical basis. *Journal of Atmospheric and Oceanic Technology* 22: 1621–1632.

Pruppacher, H., and K. Beard. 1970. A wind tunnel investigation of the internal circulation and shape of water drops falling at terminal velocity in air. *Journal of the Royal Meteorological Society* 96: 247–256.

Rosenfeld, D., and C. W. Ulbrich. 2003. Cloud microphysical properties, processes, and rainfall estimation opportunities. *American Meteorological Society Meteorological Monographs* 30: 237.

Ryzhkov, A. V. 2007. The impact of beam broadening on the quality of radar polarimetric data. *Journal of Atmospheric and Oceanic Technology* 24: 729–744.

Ryzhkov, A., S. Giangrande, and T. Schuur. 2005a. Rainfall estimation with a polarimetric prototype of WSR-88D. *Journal of Applied Meteorology* 44: 502–515.

Ryzhkov, A. V. and D. S. Zrnić. 1995. Precipitation and attenuation measurements at a 10-cm wavelength. *Journal of Applied Meteorology* 34: 2121–2134.

Ryzhkov, A. V., S. E. Giangrande, V. M. Melnikov, et al. 2005b. Calibration issues of dual-polarization radar measurements. *Journal of Atmospheric and Oceanic Technology* 22: 1138–1155.

Sachidananda, M., and D. S. Zrnić. 1986. Differential propagation phase shift and rainfall rate estimation. *Radio Science* 21: 235–247.

Sekhon, R. S., and R. C. Srivastava. 1970. Snow-size spectra and radar reflectivity. *Journal of Atmospheric Science* 27: 299–307.

Siggia, A., and R. Passarelli. 2004. Gaussian model adaptive processing (GMAP) for improved ground clutter cancellation and moment calculation. *Proceedings of 3rd European Conference on Radar in Meteorology and Hydrology*, Visby.

Straka, J. M., D. S. Zrnić, and A. V. Ryzhkov. 2000. Bulk hydrometeor classification and quantification using polarimetric radar data: Synthesis of relations. *Journal of Applied Meteorology* 39: 1341–1372.

Tokay, A., A. Kruger, and W. F. Krajewski. 2001. Comparison of drop size distribution measurements by impact and optical disdrometers. *Journal of Applied Meteorology* 40: 2083–2097.

Torlaschi, E., R. G. Humphries, and B. L. Barge. 1984. Circular polarization for precipitation measurement. *Radio Science* 19: 193–200.

Torres, S. M., and D. S. Zrnić. 1999. Ground clutter canceling with a regression filter. *Journal of Atmospheric and Oceanic Technology* 16: 1364–1372.

Ulbirch, C. 1983. Natural variations in the analytical form of the raindrop size distribution. *Journal of Climate and Applied Meteorology* 22: 1764–1775.

Vivekanandan, J., S. M. Ellis, R. Oye, et al. 1999. Cloud microphysics retrieval using S-band dual-polarization radar measurements. *Bulletin of the American Meteorological Society* 80: 381–388.

Vivekanandan, J., G. Zhang, S. Ellis, et al. 2003. Radar reflectivity calibration using differential propagation phase measurement. *Radio Science* 38: 8049.

Xue, M., M. Tong, and G. Zhang. 2009. Simultaneous state estimation and attenuation correction for thunderstorms with radar data using an ensemble Kalman filter: Tests with simulated data. *Journal of the Royal Meteorological Society* 135: 1409–1423.

Zhang, G., J. Vivekanandan, and E. Brandes. 2001. A method for estimating rain rate and drop size distribution from polarimetric radar. *IEEE Transactions on Geoscience and Remote Sensing* 39: 830–840.

Zrnić, D. S., V. M. Melnikov, and J. K. Carter. 2006. Calibrating differential reflectivity on the WSR-88D. *Journal of Atmospheric and Oceanic Technology* 23: 944–951.

4

Multi-Radar Multi-Sensor (MRMS) Algorithm

The advent of faster data transmission with Internet-2 and effective data compression techniques has enabled base-level (i.e., level II) radar data from the NEXRAD network to be transmitted and processed in real time. The first demonstration of this processing took place with the Collaborative Radar Acquisition Field Test (CRAFT) Project (Droegemeier et al. 2002; Kelleher et al. 2007). Radar data were first transmitted to regional hubs and then to a national processing and archiving center within the National Weather Service (Crum et al. 2003). One of the first hubs was in Phoenix, Arizona, comprising five surrounding WSR-88D sites. Gourley et al. (2001, 2002) demonstrated the first real-time precipitation estimation algorithm called QPE SUMS (Quantitative Precipitation Estimation and Segregation Using Multiple Sensors) that operated from the base-level radar data. Precipitation type and accumulation products were generated and made available to forecasters and water managers at the Salt River Project in Phoenix for operational use. As additional WSR-88D radars joined the network, improvements to the radar algorithms proceeded. Eventually, all radars were connected to the network and the first national QPE and radar-based products were generated in real time at the National Severe Storms Laboratory beginning in 2006. The state-of-the-science of this National Mosaic and QPE system (NMQ) using single-polarization radar data is described in Zhang et al. (2011). The entire NEXRAD network underwent an upgrade to dual-polarization technology in 2013. Moreover, the National Weather Service decided to operationalize the NMQ system beginning in 2014. A name change to Multi-Radar Multi-Sensor algorithm (MRMS) followed, corresponding to the National Weather Service operationalization and to algorithmic changes following the dual-polarization upgrade. This chapter provides a general overview of the entire MRMS algorithm.

The MRMS algorithm begins by ingesting the level II moment data (i.e., raw radar variables) from approximately 146 WSR-88D, 30 Canadian, 2 Terminal Doppler Weather Radar (TDWR), and 1 television station weather radar (KPIX). Figure 4.1 shows the locations overlaid with the three- or four-letter identifiers for each of the radars across North America. The data ingest process also incorporates approximately 9000 hourly rain gauges comprising the Hydrometeorological Automated Data System explained in detail here: http://www.nws.noaa.gov/oh/hads/WhatIsHADS.html. These data are used in QPE

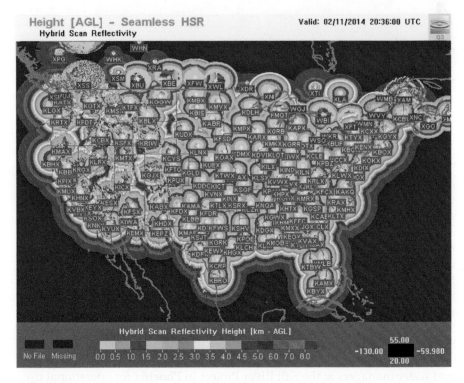

FIGURE 4.1
Hybrid scan reflectivity height product (in km above ground level); 3- and 4-letter radar identifiers are overlaid at each of the 146 WSR-88D sites, all of which contribute to the MRMS product suite.

processing for bias adjustment. Three-dimensional temperature analyses from the Rapid Refresh model (RAP) are also brought in to help guide the multisensor algorithms at several stages in the product generation. After the data are ingested into the system, the data processing begins for each individual radar. Next, mosaics are created in 2-D and 3-D space. Finally QPE products are generated and evaluated online. The most distinguishing characteristics of the MRMS products are their accuracy, national consistency, and most importantly, their resolution. The QPE and mosaic products are generated every 2 min on a 0.01 deg resolution grid, or about 1 km. The distinguishing characteristic of high resolution has gained the attraction of the satellite remote-sensing community. The MRMS rain rate products are below the pixel resolution of most satellite QPE products, even from instruments aboard low Earth-orbiting platforms. Instead of relying on statistical downscaling of rainfall products, the MRMS QPE products can be sampled up to the resolution of the satellite pixels, and accordingly, have provided invaluable information to the remote-sensing community (Kirstetter et al. 2012, 2013).

4.1 Single-Radar Processing

The single-radar process begins by performing quality control (QC) on the polarimetric radar variables in their native, spherical coordinate systems, centered on each radar. The intent of the dual-polarization QC routine is to remove all nonweather scatterers while retaining very light precipitation. If the radar is operating in clear air mode (e.g., volume coverage pattern 32) during the warm season, then all echoes are removed. Clear-air VCPs are employed with a long pulse maximize sensitivity. Thresholds are automatically monitored at the radar site so that the radar will immediately switch to precipitation mode when weather echoes have developed within the surveillance region of the radar. The removal of echoes when the radar is in clear air mode during the warm season minimizes false accumulations that would aggregate over time due to very light, but frequently occurring clear air echoes backscattered from insects and birds. Next, data are removed at bins that are deemed to have significant blockage (>50%) or the bottom of the beam does not clear the underlying terrain by at least 50 m. This step assumes the radar beam propagates as it would in a standard atmosphere relative to an accurate, underlying digital elevation model (DEM). Nonstandard beam propagation and ground-based features not represented in the DEM (e.g., trees and buildings) are dealt with in later processing steps.

4.1.1 Dual-Polarization Quality Control

The details of the dual-polarization quality control (dpQC) scheme used in MRMS are provided in Tang et al. (2014). The dpQC algorithm is based on decision-tree logic largely involving the co-polar correlation coefficient (ρ_{hv}). Before applying a ρ_{hv} threshold, the algorithm checks for hydrometeors that may be associated with intrinsically low ρ_{hv} values such as in hail, nonuniform beam filling (NUBF) situations, and the melting layer. Echoes that are not subject to the dpQC are those that have ρ_{hv} values < 0.95, the 18 dBZ echo top is higher than 8 km, and there is Z_H > 45 dBZ in the column. In general, bins meeting these criteria are tall, intense cells that may contain hail. Situations with NUBF are detected by bins with ρ_{hv} values < 0.95, the 0 dBZ echo top is higher than 9 km, and there is a cell between the bin in question and the radar with Z_H > 45 dBZ. This can cause an NUBF situation and result in depressed ρ_{hv} values at further ranges. Next, the dpQC algorithm searches for the melting layer in a manner similar to that described in Giangrande et al. (2008). There is generally a reduction in ρ_{hv} there, often in a ring around the radar, and the environmental temperature must be near 0°C. The dpQC disregards data in the suspected melting layer regime.

After those weather echoes with intrinsically low ρ_{hv} values are identified and left alone, remaining echoes are screened if they have ρ_{hv} < 0.95. But, before they are utlimately removed, the dpQC algorithm recognizes that noisy but occasionally high values of ρ_{hv} can be associated with nonweather echoes. Therefore, the algorithm examines the spatial gradient or texture of ρ_{hv}. If the standard deviation within a 1 km radial segment is less than 0.1, then the echoes are retained and subjected to the subsequent processing steps. Next, a **spike filter** is applied to identify reflectivity values that extend along small wedge-shaped radials. These spikes result from electronic interference and with a rising and setting sun. The algorithm searches for a bundle of adjacent radials with $Z_H > 0$ dBZ extending more than 30 km in range. If the number of potentially contaminated bins decreases by more than 90% when examining Z_H along the radials at the next highest tilt, then they are presumed to be from electronic interference or radiation from the sun and are subsequently removed. A spatial continuity test is applied for those bins that may have randomly high ρ_{hv} values, but are isolated. The vertical gradient test removes echoes that decrease by more than 50 dB km^{-1}. Noisy data are identified by examining the distribution of Z_H data within a 1.25 km $\times$ 1.5 deg region. If more than half the bins have missing Z_H values or the average of the nonmissing values in the neighborhood is less than 25% of the Z_H value in the center bin, then data in the center bin are deemed noisy and subsequently removed. The final cleanup step removes the entire tilt of data if the surviving bins with $Z_H >$ 10 dBZ add up to less than 10 km^2 in area.

4.1.2 Vertical Profile of Reflectivity Correction

The aims of the following processing steps are to adjust Z_H data so that they represent, as closely as possible, values that would be measured at the surface. Following the dpQC filtering steps, Z_H values are compensated in regions with partial beam blockages < 50%. Bin volumes and beam center heights increasing with range are computed assuming the 4/3 Earth's radius model and the Bessel function of second order for the power density distribution of the radar beam (Doviak and Zrnić 1993). This step results in approximately 1 dB being added for each 10% of partial beam blockage up to 50%.

Vertical profiles of reflectivity (VPRs) are constructed in spherical coordinates. This step selects data from 20 to 80 km in range from the radar and linearly interpolates data to fixed height levels spaced 200 m apart from 500 m above radar level (ARL) to 20 km. A first-pass segregation of precipitation types is conducted in spherical coordinates for convective, stratiform, and tropical echoes. This step is needed here so that averaging of stratiform profiles can take place. The convective-stratiform segregation algorithm follows the decision tree logic of Qi et al. (2013). To summarize, the partitioning assigns convective precipitation if Z_H exceeds 55 dBZ anywhere in the profile, if the vertically integrated liquid (VIL) exceeds 6.5 kgkm^{-2}, or if Z_H exceeds 35 dBZ at temperatures $<-10°$C. At this stage of processing, an algorithm based on the study of Xu et al. (2008) examines the VPR to determine if there is a "tropical" VPR that may have efficient collision-coalescence microphysics associated with warm rain processes. The tropical ID algorithm first requires that the bottom of the melting layer is at least 2 km AGL. Then, the key feature for a tropical identification is the slope of the VPR below the melting layer. If Z_h increases as the precipitation falls (negative slope of VPR), then the profile is determined to be tropical. The precipitation type flags (convective, tropical, stratiform) are stored for later use after the data have been mosaicked onto a common Cartesian grid.

A "tilt apparent VPR" is created for stratiform precipitation by taking azimuthal averages of Z_h at the four lowest tilts. Most of the precipitation typing is done in later steps after the data have been mosaicked, but some of the processing must be done in spherical coordinates and is dependent upon the approximate precipitation types. VPR corrections are made to the candidate tilts that will eventually be used to generate QPE, thus each tilt is corrected to represent surface-level equivalent Z_H. A simple, three-piece linear VPR model is fit to the azimuthally averaged VPR describing stratiform precipitation at the four lowest tilts. The three pieces correspond to (1) the pristine ice region above the top of the melting layer, (2) the top half of the melting layer (from the bright band top to the bright band peak), and (3) from the bright band peak to the surface. Given the slopes of the fit VPR models, Z_H is corrected if it is measured within the melting layer and above according to the following linear equation:

$$Z_H^{corr}(h_0) = Z_H^{obs}(h) - \left[Z_H^{VPR}(h) - Z_H^{VPR}(h_0) \right] \text{(dBZ)} \tag{4.1}$$

where the VPR superscript corresponds to the modeled reflectivity of the three-piece linear VPR, h is the height of the radar measurement, and h_0 corresponds to the bottom of the bright band, which is also approximated as the surface (assuming the Z_H values do not change appreciably below the bright band bottom). Equation (4.1) serves the purpose of reducing Z_H values measured within the bright band and increasing values that are measured aloft in the pristine ice regions. At this stage, Z_H data measured on the lowest four tilts have been corrected to represent equivalent values at the surface.

Beam blockages can become apparent in Z_H images when the beam is ducted under superrefraction conditions and/or when terrestrial features such as trees, buildings, communication towers, and wind farms are not represented in DEMs. These can cause radial streaks that have relatively low values of Z_H and are generally easily detectable by the eye. Statistically speaking, these radial streaks in Z_H generally do not impact a great volume of data, but they are generally static and can result in underestimations in resultant QPE accumulations, especially for long accumulation periods. These QPE accumulation maps are visually inspected to find these wedge-shaped regions of underestimation that emanate from the radar site. Figure 4.2 shows four of these regions

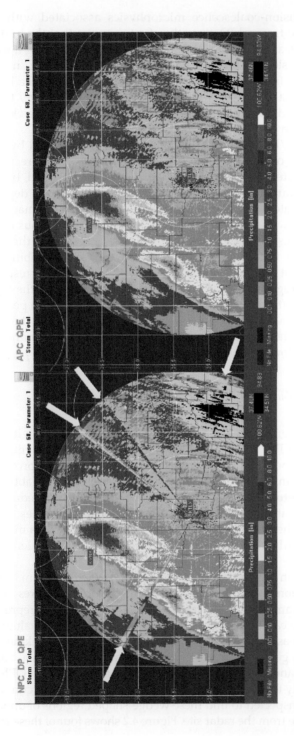

FIGURE 4.2
Examples of nonstandard beam blockages around the KTLX radar due to surrounding trees and towers that are not represented in digital elevation models. The panel on the right shows the corrected precipitation field after the nonstandard blockage mitigation procedures have been applied to remove the wedge-shaped artifacts visible along adjacent radials.

marked by yellow arrows around the KOUN radar in Norman, Oklahoma. MRMS employs a **nonstandard blockage mitigation** procedure so that Z_h data at adjacent azimuths at the same range are linearly interpolated across the artifact region to yield more seamless Z_h images and resulting QPE accumulations. Figure 4.2 shows how the interpolation procedure has predominantly mitigated the wedge shapes of underestimation. If the wedge is too large (e.g., >10 deg), then data from higher tilts are extrapolated downward to replace the region with underestimated Z_h and resultant QPEs. All of these problem regions are stored in the algorithm so that the interpolation and extrapolation procedures are applied to all future Z_h maps.

4.1.3 Product Generation

At this point, Z_h data from the lowest four elevation angles have been quality controlled, VPR-corrected, and adjusted for nonstandard beam blockage artifacts. Data from the lowest available elevation angle that clears the underlying terrain by at least 50 m and is at least 50% unblocked are used to build a 2-D **seamless hybrid scan reflectivity (SHSR)** product. There is an associated **height of SHSR (SHSRH)** product that reports the estimated height of the beam center at each range gate. The third 2-D polar product that is used in later processing is called the **Radar QPE Quality Index (RQI)** (Zhang et al. 2012). The RQI product scales from 0 to 1 and gives an indication on the expected quality of the measurement based on the degree of beam blockage and the height of the beam relative to the melting layer. It is computed as follows:

$$RQI = RQI_{blk} \times RQI_{hgt} \tag{4.2}$$

where

$$RQI_{blk} = \begin{cases} 1; blk \leq 0.1 \\ 1 - \dfrac{blk - 0.1}{0.4}; 0.1 < blk \leq 0.5 \\ 0; blk > 0.5 \end{cases} \tag{4.3}$$

and

$$RQI_{hgt} = \begin{cases} 1; h < h_{0C} - depth_{bb} \\ \exp\left[-\dfrac{\left(h - h_{0C} + depth_{bb}\right)^2}{1500^2}\right]; h \geq h_{0C} - depth_{bb} \end{cases} \tag{4.4}$$

where *blk* is the beam blockage scaling from 0 to 1, *h* is the height of the beam bottom, h_{0C} is the height of the 0°C isotherm, and *depth*$_{bb}$ is the estimated depth of the melting layer (default is 700 m). The RQI_{hgt} component essentially gives perfect values to measurements below the melting layer in the rain and then exponentially decreases the RQI values with increasing height for those measurements taken within and above the melting layer.

The next procedure mosaics the 2-D polar SHSR, SHSRH, and RQI products onto a common 2-D Cartesian grid. Recall from Section 2.6 that many procedures have been developed with varying complexity to take advantage of the fact that there can be multiple radars providing independent measurements over a given point in space. The simplest methods of merely selecting the data from the nearest radar are prone to creating linear discontinuities in resulting QPE accumulation products. To mitigate these artifacts, the MRMS algorithm employs a mosaicking scheme that is based on the height of the measurement and range from the radar. The following mosaicking logic is applied to all three SHSR, SHSRH, and RQI products.

$$f(x) = \frac{\sum_{i=1}^{N} w_r^i w_h^i x^i}{\sum_{i=1}^{N} w_r^i w_h^i} \quad (4.5)$$

where

$$w_r = \exp\left(\frac{-r^2}{L^2}\right) \quad (4.6)$$

and

$$w_h = \exp\left(\frac{-h^2}{H^2}\right) \quad (4.7)$$

f(x) is the final mosaicked product on the 2-D Cartesian grid having a 1×1 km^2 resolution, *x* represents the 2-D polar product, *i* is the radar index up to the total *N* radars contributing data at a given point, w_r corresponds to the weighting function based on the distance (*r*) between the radar and the analysis point, w_h is the weighting function based on the height of the beam center (*h*), and *H* and *L* are both shape parameters of the weighting functions. Their default values are 50 km and 1.5 km, respectively. These mosaicked data are used in subsequent steps to ultimately produce QPE products on a 1 km Cartesian grid at a 2 min frequency.

4.2 Precipitation Typology

Now that the Z_h data have been filtered and processed to create 2-D mosaics on a common Cartesian grid, the next step in MRMS involves classifying the echoes into different precipitation types. This process has been developed to deal with the different microphysical processes in precipitating clouds that result in variable DSDs. If there is DSD variability, then a single Z_h–R relation doesn't adequately apply. Forecasters in operational centers often have the capability to switch the Z_h–R parameters to accommodate different storm systems and seasons. However, the switch typically applies to all grid points underneath the radar umbrella, and thus doesn't properly apply to mixed precipitation cases such as a convective line with trailing stratiform region. Automated precipitation typing offers the advantage of incorporating detailed radar observations in three dimensions as well as multisensor data sources such as environmental temperatures and even lightning observations to aid in the decision process.

Figure 4.3 outlines the basic decision-tree logic that's used in the precipitation-typing module. The first decision implements an additional screen to remove echoes that are too weak to be associated with precipitation. Echoes with Z_H < 5 dBZ are no longer considered for QPE calculations. Echoes associated with snow are weaker due to a lower dielectric constant with frozen precipitation, therefore if the surface temperature is less than 5°C, then snow could be possible and the precipitation threshold is dropped down to 0 dBZ. Next, surface precipitation is segregated into frozen and liquid types. Note that several radar algorithms have been developed to identify different hydrometeor types and phases based on polarimetric radar observations, as was detailed in Section 3.3. It should be noted that these observations are taken at the height of the measurement and do not necessarily represent surface precipitation types. Therefore, it is very important to incorporate environmental observations to more accurately estimate surface precipitation types. The separation of frozen and liquid surface precipitation is based on two temperature thresholds. If the surface wet bulb temperature (as determined from the RAP) is less than 0°C and the surface dry bulb temperature is less than 2°C, then the surface precipitation type is set to frozen. The use of two thresholds accounts for situations in which surface temperatures are just above freezing, but wet snow is reaching the surface. The next decision identifies bin volumes dominated by hailstones using the Maximum Expected Hail Size (MESH; Witt et al. 1998) algorithm. MESH was calibrated on observed hail sizes using the Severe Hail Index, which is a vertical integral from the melting layer to storm top (in the ice region) of Z_H > 40 dBZ. The MRMS algorithm identifies hail if the MESH is nonzero.

Convective echoes are identified on a bin-by-bin basis using the same criteria that were implemented during the processing in spherical coordinates. The conditions are the 0°C height > 1.5 km, and VIL > 6.5 kg km^{-2}, or Z_H > 35 dBZ

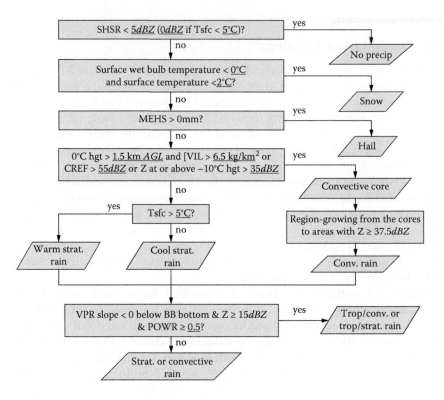

FIGURE 4.3
Overview of the decision-tree logic used in MRMS to define precipitation types. This precipitation typology is subsequently used to guide the application of Z–R equations.

at temperatures $< -10°C$, or the composite (column maximum) $Z_H > 55$ dBZ. If bins adjacent to a convective bin do not meet the aforementioned criteria but have composite $Z_H > 35$ dBZ, then they are assumed to be in growing convective regions with presumably strong updrafts and are also assigned as convective precipitation type. Nonconvective cells are further subdivided into cool and warm stratiform categories if they do not meet the convective criteria and their surface temperatures are less than or greater than 5°C, respectively. The final step in the precipitation typology is the application of the tropical identification algorithm using similar logic applied on the data in spherical coordinates. That is, the VPR slope must be negative below the melting layer and the SHSR must be greater than 15 dBZ. Additional decisions are made based on the concept of the **probability of warm rain (POWR)**.

The POWR algorithm described in Grams et al. (2014) assigns probabilities from 0 to 1 to each bin based primarily on environmental variables. The key predictors for warm rain are temperature lapse rate from 850 to 500 hPa being close to moist adiabatic, a relatively high melting layer height, and high relative humidity in low levels from 1000 to 700 hPa. These factors result in weak-moderate updrafts within a very deep, moist environment. These conditions enhance collision-coalescence processes and result in DSDs that are anomalous in that there are relatively high populations of small droplets. If the prior tropical conditions are met but the POWR is less than 0.5, then the original convective, warm stratiform, and cool stratiform assignments remain. If POWR is greater than 0.5, then the DSD is assumed to be mixed between convective or stratiform and tropical characteristics. These grid cells are assigned a mixed convective/tropical or a mixed stratiform/tropical type, which adjusts the QPE scheme described below.

4.3 Precipitation Estimation

Now that each grid cell has been assigned a precipitation type, the MRMS algorithm assigns an appropriate Z_h–R relation on a cell-by-cell basis. If the POWR is < 0.5, then the following relations are used for warm and cool stratiform, convection, hail, and snow:

$$Z_h = 200\,R^{1.6} \qquad \text{for warm stratiform} \qquad (4.8)$$

$$Z_h = 130\,R^{2.0} \qquad \text{for cool stratiform} \qquad (4.9)$$

$$Z_h = 300\,R^{1.4} \qquad \text{for convection} \qquad (4.10)$$

$$Z_h = 300\,R^{1.4} \qquad \text{for hail} \qquad (4.11)$$

$$Z_h = 75\,S^{2.0} \qquad \text{for snow} \qquad (4.12)$$

where Z_h is in mm^6 m^{-3} and R is in mm hr^{-1}. Because Z_h is very sensitive to large-diameter hydrometeors, it is common practice to place an upper limit on R values so as to avoid unrealistically large accumulations. These upper limits, or caps, are set to 48.6, 36.5, 103.8, and 53.8 mm hr^{-1} for each respective precipitation type from Equations (4.8)–(4.11). No cap is enforced for snow water equivalent estimation in Equation 4.12. Figure 4.4 shows the Z_h–R curves for each of the relations. We can see that the selection of the different curves corresponding to different precipitation types has a major impact on the estimated precipitation rates, especially for the larger reflectivity values. For instance, precipitation rates are approximately doubled when going from cool stratiform to convective precipitation type for $Z_h = 50$ dBZ.

If the POWR is ≥0.5 for grid cells west of 100°W, then the following equation is used:

$$Z_h = 250\,R^{1.2} \qquad \text{for tropical} \qquad (4.13)$$

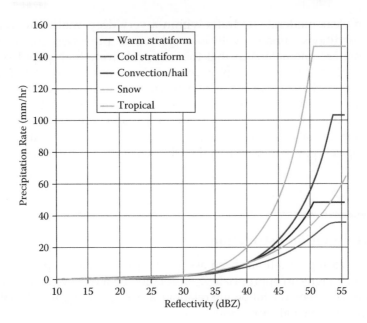

FIGURE 4.4
Reflectivity-to-precipitation rate relations used in the MRMS algorithm. Each of the precipitation types shown in the legend is automatically identified using volumetric radar data and environmental data from the RAP model analysis.

with a cap set to 147.4 mm hr^{-1}. Tighter constraints on the tropical precipitation type are applied for echoes east of 100°W. A study conducted by Chen et al. (2013) revealed overestimation by the MRMS daily precipitation accumulations, primarily in the southeastern United States. Analysis of the precipitation type assignments revealed that the overestimation was due to the tropical precipitation type being assigned too frequently, especially during the cool season when it is not expected. So, the MRMS now computes weighted rainfall rates for POWR ≥ 0.5 that are a blend between Equations (4.8)–(4.10) and (4.13) as

$$R_{mix} = \frac{w_{conv}R_{conv} + \alpha w_{trop}R_{trop}}{w_{conv} + w_{trop}} \qquad \text{for mixed type} \qquad (4.14)$$

where the weights (w_{conv}, w_{trop}) vary from 0 to 1 depending on POWR. Note that Equation (4.14) applies to the case of mixed convection and tropical precipitation type. The same logic also applies to mixes between cool stratiform and tropical, and warm stratiform and tropical. There is an additional dynamic weighting factor (α) that enables more weighting to be applied to R_{trop} from June to November, with maximum weight occurring in September. This is a physical constraint that ensures that most tropical precipitation types will be assigned during the warm season when they are expected. This is useful because the tropical Z_h–R relation yields the highest R per unit Z_h. Incorrect assignment of this precipitation type will lead to large overestimations, so it must be used with careful constraints.

The precipitation rates that are produced using the aforementioned procedures are generated every 2 min on a 1 km grid and are referred to as the radar-only product. These rainfall rates are summed to create hourly accumulations, which are also output every 2 min. Longer accumulations of 3, 6, 12, and 24 hr are created at the top of each hour from the hourly accumulation products. Once daily, 48 and 72 hr accumulations are produced at 1200 UTC. The radar-only products are most useful for those applications that require rainfall rates rather than longer term accumulations, such as for forcing a flash flood prediction model or an urban flood model. If longer accumulations are needed, then there is an opportunity to perform bias adjustment to the radar-only products using collocated rain gauges, as is described next.

MRMS ingests approximately 9000 gauges across the United States each hour and then compares them to the radar-only hourly product at the top of the latest hour. The **local gauge bias-corrected (LGC) radar** product computes the bias as $(b_k = r_k - g_k)$ at each kth gauge site where r signifies the radar-only hourly accumulation and g is the gauge value. Next, an inverse distance-weighting scheme is used to spatially interpolate b_k values onto the 2-D Cartesian grid as

$$b_i^a = \alpha \frac{\sum\limits_{k=1}^{N} w_{i,k} b_k}{\sum\limits_{k=1}^{N} w_{i,k}} \tag{4.15}$$

where

$$w_{i,k} = \begin{cases} \dfrac{1}{d_{i,k}^b} ; d_{i,k} \leq D \\[2mm] 0 ; d_{i,k} > D \end{cases} \tag{4.16}$$

and

$$\alpha = \min\left\{ \sum\limits_{k-1}^{N} \exp\left[-\frac{d_k^2}{(D/2)^2} \right]; 1.0 \right\} \tag{4.17}$$

where the d is the Euclidean distance from the ith analysis grid point and kth rain gauge. The parameters b and D in Equation (4.16) represent the shape of the weighting function and the cutoff distance, respectively. Both parameters are optimized each hour using a leave-one-out cross-validation scheme. A randomly selected gauge is purposely left out of the analysis, and values of b and D are cycled until optimum values are found for the bias that is already known. These parameter values are stored and then another gauge site (and its bias) is purposely withdrawn from the analysis to find values for b and D. This procedure is repeated until all gauges have been included in the cross-validation scheme to find optimum values for b and D, which depend on gauge density and rainfall variability. The gauge density parameter α in Equation (4.17) reduces the impact of the bias correction when the gauge density is sparse. The final step in the local gauge correction removes the spatially interpolated bias computed in Equation (4.15) from the radar-only hourly accumulations.

Two more gauge-based products complete the suite of QPE products in MRMS. The **gauge-only** product ingests hourly accumulations from the same gauges used to build the LGC product described above. The same leave-one-out cross-validation scheme used to spatially interpolate the bias is used here to interpolate the gauge values themselves in the gauge-only product. The parameters b and D are optimized specifically for the gauge-only product and a gauge-based accumulation is produced at every grid point. The second gauge-based product is similar to the gauge-only product, but utilizes monthly precipitation climatologies from the PRISM (Parameter-elevation Relationships on Independent Slopes Model) dataset described in Daly et al. (1994). For the **mountain mapper QPE**, the bias is computed as $(b_k = g_k/p_k)$ where g_k is the hourly gauge accumulation and p_k is the monthly climatological value. The bias is then interpolated onto the 2-D Cartesian grid using the inverse distance weighting scheme in Equations (4.15) and (4.16) with $\alpha = 1$. The mountain mapper QPE essentially adjusts the known precipitation climatology in mountainous terrain to the observed gauge amounts. This technique has proven to be useful in regions where the spatial patterns of precipitation are heavily dictated by the underlying terrain due to orographic effects. Caution must be exercised, however, for extreme events when the precipitation patterns may shift from their climatological locations due to anomalous wind directions, speeds, and/or water vapor contents. The gauge-based hourly products are generated at the top of each hour. Accumulations of 3, 6, 12, and 24 hr are computed at the top of each hour from the hourly products. Longer term accumulations of 48 and 72 hr, derived from the 24 hr accumulations, are computed at 1200 UTC each day.

4.4 Verification

Rain gauges serve as inputs to several of the QPE products, and they are also instrumental in evaluating and further improving the radar-only product. Recall that the 2 min/1 km resolution with this product is needed for several satellite QPE and flash flood applications. Further, improvements to the radar-only algorithm cascade to the local gauge bias-corrected radar product and are very important in regions that do not have dense gauge coverage. Evaluation with rain gauges is useful in the event that they are independent from the QPE algorithm. Moreover, they must have high quality to serve as the reference or "ground truth" dataset. As we will see in this section, gauge accumulations can have errors of their own and must be quality controlled as well.

MRMS employs a robust QPE verification system that automatically ingests automated rain gauge reports and compares them to user-selectable MRMS QPE products. Products are evaluated using (1) statistics, (2) gauge circle plots, and (3) scatterplots. Figure 4.5 shows a 24 hr accumulation ending 0000 UTC on October 15, 2013, from the radar-only algorithm in the background with the gauge circles overlain. The diameter of each circle corresponds to the gauge accumulation and the color represents the bias shown in the color table. This plot is useful for showing the geographic dependence of the biases. For example, we see the greatest underestimation with the largest accumulations in the southern part of the image. Moreover, the plot reveals overestimation for very light rainfall accumulations on the periphery of the rain system to the northeast. The scatterplot in Figure 4.6 plots the radar-only

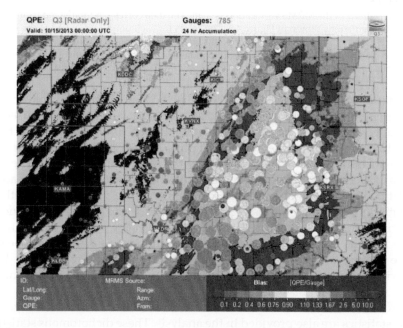

FIGURE 4.5
Gauge circle plot used in the MRMS verification system. The gridded product is the radar-only, 24 hr accumulation ending on October 15, 2013. The circles are centered on the gauge locations with their radii proportional to the gauge accumulations. The colors of the circles correspond to biases as shown in the legend.

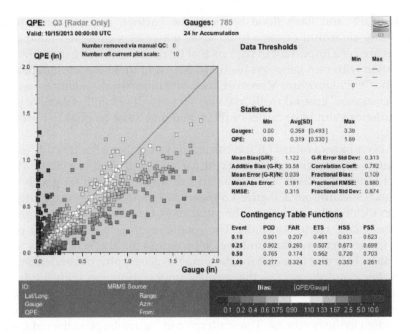

FIGURE 4.6

Scatterplot used in the MRMS verification system. The data correspond to the radar-only and gauge accumulations shown in Figure 4.5. The colors for each point correspond to the bias as in Figure 4.5. Notice the large number of points on the y-axis corresponding to nonzero radar-only accumulations with zero gauge accumulations. Commonly used statistics are shown in the right side of the figure panel.

accumulations against the collocated gauge accumulations and color codes each point based on the individual bias using the same color scale used in the gauge circle plots. We can see that for gauge accumulations greater than 0.5″ (12.7 mm), most of the points are falling to the right of the 1:1 diagonal line, which indicates underestimation by the radar-only algorithm. There is also a collection of purple-colored points on the ordinate indicating no accumulation by rain gauges, but nonzero amounts by the radar-only algorithm.

Statistics are shown on the right side of Figure 4.6. Values are provided describing the distribution of the radar-only and gauge-based accumulations including the minimum, maximum, mean, and standard deviation. Below those values, a suite of continuous variable statistics are computed to describe the discrepancies between the radar-only and gauge accumulations ranging from bias (multiplicative and additive), mean absolute error, and root-mean-squared error, to correlation coefficient. Several contingency table statistics are also provided in the analysis. These dichotomous statistics differ from the continuous variable ones in that they indicate whether an event did or did not occur. The "events" are based on four thresholds applied to the accumulated rain gauge totals. The values shown correspond to the probability of detection (POD; event was forecast and it occurred), false alarm rate (FAR; event was forecast but did not occur), and different skill metrics that combine several variables in the contingency table along with climatological information.

The multiplicative bias of 1.122 implies that gauges were accumulating approximately 12% more rain than the radar-only algorithm, which is generally considered a good result. Figure 4.7 uses a capability in the QPE verification system to choose the specific network of rain gauges. In this case, only rain gauges comprising the Oklahoma Mesonet are used for the evaluation. Note that the Mesonet gauges undergo rigorous maintenance in the field and their data are quality controlled using established automated procedures as well as manual checks. When only the high-quality Mesonet gauges are used in the scatterplot we see that the initial conclusion of underestimation by the radar-only product is further supported, and the points corresponding to nonzero radar-only amounts but zero gauge accumulation have been removed. Most importantly, the original multiplicative bias of 1.122 has now grown to 1.438. The selection of high-quality gauges removed all the points residing on the ordinate, which were counterbalancing the true underestimation that was present in the radar-only algorithm. This example highlights the need for careful consideration of the gauge quality when conducting evaluations of QPE algorithms.

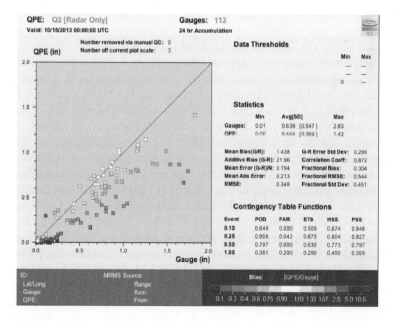

FIGURE 4.7
Same as in Figure 4.6, but the data in this figure come from high-quality rain gauge data in the Oklahoma Mesonet. While the nonzero/zero radar/gauge pairs have disappeared, the analysis indicates a high bias by the radar-only algorithm. The statistics on the right also show this.

4.5 Discussion

Despite the advances in the state-of-the-science of QPE represented in MRMS, the present algorithm makes little direct use of polarimetric variables in precipitation estimation. Future research will focus on systematic QPE using the polarimetric variables given that they offer the potential to respond to different DSDs. This is the main objective in MRMS's current approach of identifying different precipitation types and applying different Z_h–R relations, as detailed in Section 4.2. It is presumed that the different vertical structures of Z_h as a function of temperature highlight different microphysical processes that result in DSD variability. Polarimetric rainfall relations are currently running on the operational system in the National Weather Service as single radar products. Additional work is needed to determine optimum strategies to mosaic the polarimetric variables and/or rainfall estimators. Furthermore, prior studies performed on the polarimetric rainfall estimators have been done on individual radars over limited geographic domains. Application of a given algorithm to the entire network of NEXRAD radars needs to be explored to properly assess and improve the parameters.

The lack of NEXRAD radar coverage at low altitudes in the Intermountain West is a problem that has not been fully mitigated by MRMS. The mountain mapper technique partially addresses this through the use of rain gauge accumulations that are spatially interpolated based on monthly climatologies. This method depends on the presence of well-maintained rain gauge networks as well as the true precipitation patterns remaining similar to their climatological spatial characteristics. This latter assumption can fail during extreme events, which are anomalous and deviate from climatological conditions. Another approach that is being explored by the MRMS development team is the integration of passive and active sensors from spaceborne platforms. These sensors have the unparalleled advantage of being able to look down on the precipitating systems and thus are less impacted by intervening terrain. The disadvantages of satellite-based approaches are the indirectness of the signals as they relate to surface precipitation rates and the frequency at which they provide information. Active radar sensors, which provide reflectivity data just like ground radars, must be aboard low Earth-orbiting satellites in order to have a reasonable small pixel resolution. This means that an overpass at a given location is only available once or twice a day. Passive sensors aboard geostationary satellites can provide data at high spatiotemporal resolutions similar to the ground radars, but their radiance signals (i.e., brightness temperatures at cloud top) are indirectly related to surface precipitation rates. Nonetheless, detailed vertical profiles of reflectivity from low Earth-orbiting active sensors can be combined with the passive signals to fill in the gaps present with the NEXRAD radar network.

Problem Sets

Q1: List the products from the MRMS system along with their spatial and temporal resolutions.

Q2: Compute and plot rain rates using the three typical relationships given below.

a. $Z = 200\ R^{1.6}$ (Marshall–Palmer Z–R)

b. $Z = 300\ R^{1.4}$ (WSR-88D conventional Z–R)

c. $Z = 250\ R^{1.2}$ (WSR-88D tropical Z–R)

What is the percent different in computed rainfall rates from the different relations for a Z value of 20 dBZ and 50 dBZ? List some reasons why the different Z–R relations exist.

References

Chen, S., J. J. Gourley, Y. Hong, P. E. Kirstetter, J. Zhang, K. W. Howard, Z. L. Flamig, J. Hu, and Y. Qi. 2013. Evaluation and uncertainty estimation of NOAA/NSSL next-generation National Mosaic Quantitative Precipitation Estimation product (Q2) over the continental United States. *Journal of Hydrometeorology* 14: 1308–1322. doi:10.1175/JHM-D-12-0150.1.

Crum, T. D., D. Evancho, C. Horvat, M. Istok, and W. Blanchard. 2003. An update on NEXRAD program plans for collecting and distributing WSR-88D base data in near real time. Preprints, *19th International Conference on Interactive Information Processing Systems (IIPS) for Meteorology, Oceanography, and Hydrology,* February 9–13, Amer. Meteor. Soc., Long Beach, CA, Paper 14.2.

Daly, C., R. P. Neilson, and D. L. Phillips. 1994. A statistical-topographic model for mapping climatological precipitation over mountainous terrain. *Journal of Applied Meteorology* 33: 140–158.

Doviak, R. J. and D. S. Zrnić. 1993. *Doppler Radar and Weather Observations.* Academic Press, San Diego, 562 pp.

Droegemeier, K. K., et al. 2002. Project CRAFT: A test bed for demonstrating the real time acquisition and archival of WSR-88D Level II data. Preprints, *18th International Conference on Interactive Information Processing Systems (IIPS) for Meteorology, Oceanography, and Hydrology,* January 13–17, Amer. Meteor. Soc., Orlando, FL, 136–139.

Giangrade, S. E., J. M. Krause, and A. V. Ryzhkov. 2008. Automatic designation of the melting layer with a polarimetric prototype of the WSR-88D radar. *Journal of Applied Meteorology* 47: 1354–1364.

Gourley, J. J., J. Zhang, R. A. Maddox, C. M. Calvert, and K. W. Howard. 2001. A real-time precipitation monitoring algorithm: Quantitative Precipitation Estimation and Segregation Using Multiple Sensors (QPE SUMS). Preprints *Symp. on Precipitation Extremes: Prediction, Impacts, and Responses,* Albuquerque, Amer. Meteor. Soc., 57–60.

Gourley, J. J., R. A. Maddox, K. W. Howard, and D. W. Burgess. 2002. An exploratory multisensor technique for quantitative estimation of stratiform rainfall. *Journal of Hydrometeorology* 3: 166–180.

Grams, H., J. Zhang, and K. Elmore. 2014. Automated identification of enhanced rainfall rates using the near-storm environment for radar precipitation estimates. *Journal of Hydrometeorology* 15(3): 1238–1254.

Kelleher, K. E., et al. 2007. Project CRAFT: A real-time delivery system for NEXRAD level II data via the Internet. *Bulletin of the American Meteorological Society* 88: 1045–1057.

Kirstetter, P.-E., Y. Hong, J. J. Gourley, S. Chen, Z. L. Flamig, J. Zhang, M. Schwaller, W. Petersen, and E. Amitai. 2012. Toward a framework for systematic error modeling of spaceborne precipitation radar with NOAA/NSSL ground radar-based National Mosaic QPE. *Journal of Hydrometeorology* 13: 1285–1300. doi:10.1175/JHM-D-11-0139.1.

Kirstetter, P.-E., Y. Hong, J. J. Gourley, M. Schwaller, W. Petersen, and J. Zhang. 2013. Comparison of TRMM 2A25 products, version 6 and version 7, with NOAA/NSSL ground radar-based National Mosaic QPE. *Journal of Hydrometeorology* 14: 661–669. doi:10.1175/JHM-D-12-030.1.

Qi, Y. , J. Zhang, P. Zhang. 2013. A real-time automated convective and stratiform precipitation segregation algorithm in native radar coordinates. *Journal of the Royal Meteorological Society* 139: 2233–2240.

Tang, L., J. Zhang, C. Langston, J. Krause, K.Howard, and V. Lakshmanan. 2014. A physically based precipitation/nonprecipitation radar echo classifier using polarimetric and environmental data in a real-time national system. *Weather and Forecasting*. In press.

Witt, A., M. D. Eilts, G. J. Stumpf, J. T. Johnson, E. D. W. Mitchell, and K. W. Thomas. 1998. An enhanced hail detection algorithm for the WSR-88D. *Weather and Forecasting* 13: 286–303.

Xu, X., K. Howard, and J. Zhang. 2008. An automated radar technique for the identification of tropical precipitation. *Journal of Hydrometeorology* 9: 885–902.

Zhang, J., K Howard, et al. 2011. National mosaic and multi-sensor QPE (NMQ) system: Description, results and future plans. *Bulletin of the American Meteorological Society* 92: 1321–1338.

Zhang, J., K. Howard, C. Langston, and B. Kaney. 2012. Radar Quality Index (RQI): A combined measure of beam blockage and VPR effects in a national network. *Proceedings International Symposium on Weather Radar and Hydrology,* IAHS Publ. 351–25 (2012).

5

Advanced Radar Technologies for Quantitative Precipitation Estimation

Quantitative precipitation estimates (QPEs) are conventionally provided by ground radar networks. Ground-based systems are able to acquire data routinely at relatively low altitudes up to ranges on the order of 300 km, provided there are no appreciable blockages by intervening terrain. As will be detailed in this chapter, it is also feasible to obtain radar measurements from space. Spaceborne measurements offer the advantages of collecting data over oceanic and data-sparse regions without the constraints of international borders or beam blockages by mountains. Spaceborne systems cannot measure precipitation as frequently as ground radar systems due to orbital restrictions with low Earth-orbiting satellites. They also have limitations with ground clutter contamination, nonuniform beam filling, and attenuation. Nonetheless, great potential exists to synergize radar measurements from ground and space so as to fill in voids in the operational ground radar networks.

In recent decades, we have witnessed great technological advances in radar design. Costs have also decreased, making it feasible to build small, portable radars for use in research and operations. Because the radar beam volume increases with range, it is often advantageous to collect data at close range for detailed studies of tornadogenesis and microphysical processes. Some X-, C-, Ku-, Ka-, and W-band radars can be mounted on a flatbed truck and transported directly to the location where a certain event takes place. These radars are called **mobile radars**. Mobile radar systems offer flexibility in deployment that extends or supplements the operational coverage by filling in gaps and increases the likelihood of sampling a particular type of event. In some instances an operational radar failed and was temporarily replaced by a mobile radar that was driven to a site near the operational WSR-88D radar. These radars can also operate continuously with rapid sampling with a focus on the lowest altitudes of storm systems (Biggerstaff et al. 2005). The following sections introduce some examples of advanced radar systems, ranging from small mobile radars, single and dual-frequency spaceborne radar systems, and phased-array polarimetric radars for rapid scanning of storms. This is not meant to be a comprehensive discussion of all mobile radars, as there are many throughout the world, but rather a description of a diverse selection of contemporary radar system.

5.1 Mobile and Gap-Filling Radars

Radars can operate at different wavelengths or radio frequencies. The details of the most typical radar wavelengths employed for hydrological uses are provided in Table 1.1. Longer wavelength radars like S-band are less prone to attenuation by precipitation, but they require a larger, heavier antenna, which requires a powerful pedestal, and have higher associated costs. Therefore, the most typical radar frequencies used for mobile radars range from C- to W-band. C-band radars are used commonly for operational surveillance in Europe and Canada. The signal is not as prone to attenuation loss as at X-band, but a larger antenna is needed. X-band has become more popular for mobile radars because of the smaller antenna and smaller beamwidth that can be used. Moreover, it turns out that the polarimetric variables are more predictable in the Rayleigh scattering regime at X-band than at C-band. This means that the variables increase more monotonically with larger drops, which is not the case at C-band. X-band radars have more limited ranges, which can be compensated by moving the radar close to the meteorological phenomena of interest. Below, we introduce several C- and X-band radars that operate on mobile platforms for hydrological purposes, including the SMART-R, AIR, NOXP, and PX-1000. All these radars were designed and maintained in the National Weather Center in Norman, Oklahoma, by the Advanced Radar Research Center (ARRC) of the University of Oklahoma or the National Severe Storms Laboratory (NSSL).

5.1.1 ARRC's Shared Mobile Atmospheric Research and Teaching Radar (SMART-R)

The SMART-R is one of the earliest mobile radars having a legacy from the Dopper-on-Wheels (DOW). The first SMART-R (SR-1) was developed and used by a consortium of scientists and engineers from the University of Oklahoma, NSSL, Texas A&M University, and Texas Tech University. They have been deployed during a number of field experiments for storm-scale research and to enhance graduate and undergraduate education in radar meteorology (Biggerstaff et al. 2005). The lower C-band frequency reduces signal loss from attenuation and improves the Nyquist (unambiguous) velocity that can be measured with the radar. The tradeoff is the resolution for observing small-scale circulations, because C-band radars have a larger beamwidth than X-band radars in terms of specified antenna size. The characteristics of the SMART-Rs are described in Table 5.1. Note that the second SMART-R (SR-2) operates at a slightly different frequency than the first SMART-R (SR-1), and SR-2 has been upgraded with dual-polarization capability.

The SMART-Rs have contributed to a great deal of research and education. It has been deployed in a series of field projects to study atmospheric phenomena ranging from tornadogenesis, hurricanes, rain in the tropics, to cool season orographic rain and snow. These radars are especially unique in that they

TABLE 5.1

Characteristics of SMART Radars

Subsystems	Description
Platform	
Physical dimensions	4700 International dual-cab diesel truck
	Length ~10 m (32 ft. 10 in.); height ~4.1 m (13 ft. 6 in.); weight ~11,800 kg total system
Power plant	10 kW diesel generator
Leveling system	Computer assisted; variable rate manual hydraulic controls
Transmitter	
Frequency	5635 MHz (SR-1), 5612.82 MHz (SR-2)
Type	Magnetron; solid-state modulator and high-voltage power supply
Peak power	250 kW
Duty cycle	0.001
Pulse duration	Four predefined values selectable from 0.2 to 2.0 μs
Polarization	Linear horizontal (SR-1); Dual linear, SHV (SR-2)
Antenna	
Size	2.54 m diameter solid parabolic reflector
Gain	40 dB (estimated)
Half-power beam	Circular, 1.5 deg wide
Rotation rate	Selectable from 0 to 33 deg s^{-1}
Elevation range	Selectable from 0 to 90 deg
Operational modes	Pointing, full PPI, range–height indicator (RHI), sector scans
Signal processor	SIGMET
Maximum number of bins per ray	2048
Bin spacing	Selectable from 66.7 to 2000 m
Moments	Radar reflectivity (filtered and unfiltered), velocity, spectrum width
Ground clutter filter	Seven user-selectable levels
Range averaging	Selectable
Dual-pulse repetition frequency de-aliasing	Selectable
Processing modes	Pulse pair, fast Fourier transform, random phase
Data archive	CD-ROM; SIGMET IRIS format
Display	Real-time PPI ; loop, pan, and zoom PPI or RHI products

Source: Adapted from Biggerstaff et al. (2005).

have been used in undergraduate and graduate radar meteorology classes to give students opportunities to obtain practical experience operating weather radars. The SMART-Rs has proven to be effective in obtaining high temporal and spatial resolution data over mesoscale domains, which is useful for basin-scale hydrological studies. Gourley et al. (2009) deployed SR-1 to the American River Basin near Sacramento, California, during the cool season of 2005–2006 during the Hydrometeorological Testbed (HMT) experiment. They examined the impact of incremental improvements to QPE processing, including calibrating Z with a nearby disdrometer, optimizing parameters in the Z–R relation, VPR correction, and maximizing low-level coverage by merging radar data with another mobile radar nearby. Their study quantified the improvements following each step. They also highlighted challenges in siting a mobile radar in complex terrain. Blockages from nearby trees, which don't appear in a digital elevation model, prevented low-altitude coverage over upper parts of the basin.

5.1.2 NSSL's X-Band Polarimetric Mobile Radar (NOXP)

NSSL's X-band Polarized Mobile Radar (NOXP) was built on a flatbed international truck frame by a group of NSSL engineers and technicians in 2008 (Palmer et al. 2009). NOXP is a dual-polarization research radar that is basically a clone of SR-2 but operates at X-band. Detailed characteristics of the radar are provided in Table 5.2. It has been used to study tornadogenesis during the Verification of the Origins of Rotation in Tornadoes Experiment–II (VORTEX-II) in 2009. It was also deployed to desert regions in Arizona during the summers of 2012 and 2013 to study thunderstorms, microbursts, and dust storms (haboobs). NOXP was shipped to France for the hydrological cycle in the Mediterranean Experiment during the autumn of 2012 (HyMeX; Ducrocq et al. 2014). NOXP observed intense precipitation rates that often produce flash flooding in the Cevennes–Vivarais region in the south of France. Several other ground instruments nearby have been used synergistically to study cloud microphysical processes, cloud electrification, and quantifying the impact of Mediterranean moisture transport on precipitating systems (Bousquet et al. 2014). NOXP observed a Z_{DR} dipole that was also associated with decreasing Φ_{dp} values in upper levels of a thunderstorm. It turns out these artifacts were actually signatures of depolarization caused by ice particles that had become oriented in the same direction due to a strong electrical field. This hypothesis was later tested and supported using lightning mapping array (LMA) observations that were deployed for the experiment.

In 2014, NOXP was deployed to the Smoky Mountains in North Carolina during the Integrated Precipitation and Hydrology Experiment (IPHEx) (Figure 5.1). Radar data collection was coordinated with NASA's S-band, polarimetric (NPOL) radar, as well as aircraft flying in and above the clouds. The radar was positioned on a mountaintop with an unimpeded view in the southern quadrant over the Pigeon River basin. Note how the hydraulic supports were used on the steep slope to properly level the antenna. The NOXP radar provided

TABLE 5.2

Characteristics of the NSSL's X-Band Polarimetric Mobile Radar

Parameter	Value
General	
Wavelength	3.22 cm
Mobile/transportable/fixed	Mobile (0.88 deg half-power beamwidth)
Scanning/profiler	Scanning (1.0 deg resolution)
Conventional/Doppler/polarimetric	Dual-polarimetric (STaR) and H-only mode
Scan capabilities	5 rpm (30 deg/s) in azimuth; 0–91 deg elevation; RHI capable
Range	Max range defined by selectable PRF; previous deployment used 1350 pulses/s, which equates to 111 km
Transmitter/receiver	
Frequency (MHz)	9410
Peak power at antenna port (dBW)	47
Equivalent isotropically radiated power (EIRP) (dBW)	47
Modulation type	Pulse
Characteristics of the modulation (e.g., sweep period, sweep rate)	None
Spectrum width pattern (MHz)	at –3 dB : 4
Antenna	
Antenna type	Dish
Antenna gain (dBi)	45.5
–3 dB antenna aperture (°)	0.9
Relative gain at horizon (dBi)	45.5
Polarization	Dual Linear
Rotation speed (rpm) (min and max)	0–5

low-level coverage, much better than what is available using NEXRAD, over three gauged basins with catchments less than 150 km^2. This provided a unique opportunity to study streamflow response to rainfall (estimated by different platforms) in small basins in complex terrain prone to flash flooding.

5.1.3 ARRC's Atmospheric Imaging Radar (AIR)

The AIR is another small radar system developed by the ARRC that employs imaging technology to simultaneously gather volumetric data on a mobile platform (Isom et al. 2013). The radar is mounted on the back of a radar truck and is thus fully mobile (Figure 5.2). The AIR doesn't use a conventional parabolic antenna that transmits and receives the electromagnetic signal. The AIR consists of 36 separate subarrays on a flat panel and is able to create an array of beams along a single dimension. It uses digital beam-forming (DBF)

FIGURE 5.1
The National Severe Storms Laboratory's NOXP mobile, polarimetric, X-band radar. Here, it is operating in the Pigeon River basin in the Smoky Mountains of western North Carolina during the Integrated Precipitation and Hydrology Experiment (IPHEx) in 2014.

FIGURE 5.2
The University of Oklahoma Advanced Radar Research Center's Atmospheric Imaging Radar (AIR). This mobile radar does not use a conventional parabolic dish antenna, but transmits a wide (1 × 20 deg) fan beam and receives the backscattered signal using 36 independent subarrays.

TABLE 5.3

Characteristics of the ARRC's Atmospheric
Imaging Radar (AIR)

Parameter	Value
General	
Frequency	9.55 GHz
Power	3.5 kW TWT
Duty cycle	2%
Sensitivity	10 dBZ at 10 km
Range resolution	30 m (pulse compression)
Subarrays	
3 dB beamwidth	1 × 20 deg
Gain	28.5 dBi
VSWR	2:1
Polarization	Horizontal (RHI mode)
Array	
Beamwidth	1 × 1 deg
Number of subarrays	36
Pedestal	
Rotation rate	20 deg s^{-1}

technology to steer the beam electronically, which results in very high temporal resolution of the collected data. The concept is to essentially transmit a single, fan beam (1 × 20 deg) and then receive individual components using the 36 independent subarrays. The beam forming is done in postprocessing using the DBF concept. The AIR operates at X-band, providing a balance between sensitivity, attenuation, and physical size, and is used as precipitation radar with a primary focus on rapidly evolving weather phenomena, namely severe storms. With range-height indicator (RHI) mode as the primary operation mode, DBF will occur in the vertical dimension, which provides near-continuous coverage of the atmosphere over a 20 deg field of view. A summary of the characteristics of the AIR is given in Table 5.3.

5.1.4 ARRC's Polarimetric X-Band 1000 (PX-1000)

The PX-1000 is a traveling wave tube (TWT)-based, transportable, dual-polarization X-band radar developed at the ARRC (Cheong et al. 2011). The system features a pair of 1.5 kW TWT transmitters, a 1.2 m parabolic reflector dish with dual-polarization feed and an azimuth-over-elevation pedestal. The general system characteristics of the PX-1000 are described in Table 5.4. PX-1000 uses a unique signal processor for complex operations such as pulse compression, multilag moment estimation, and a nonlinear frequency modulator. The PX-1000 is mounted on a trailer, which makes it transportable

TABLE 5.4

Characteristics of the ARRC's PX-1000
Transportable, X-Band, Polarimetric Radar

Parameter	Value
General	
Operating frequency	9550 MHz
Sensitivity	7 dBZ @ 50 km
Estimated system loss	2 dB
Observation range	>60 km
Transmitter	
Peak power	1.5 kW
Maximum pulse width	15 us
Maximum duty cycle	2%
Antenna	
Antenna gain	38.5 dBi
Diameter	1.2 m
3 dB beamwidth	1.8 deg
Polarization	Dual linear

rather than fully mobile (Figure 5.3). The design of the radar is well suited for long deployments in fixed locations, such as throughout seasonal field campaigns.

5.1.5 Collaborative Adaptive Sensing of the Atmosphere (CASA)

As we saw in Chapter 2, the NEXRAD radar network has limited low-altitude coverage for regions within the Rocky and Sierra Mountains. One possible solution to mitigate these radar data voids is to increase the number of radars within the data-sparse regions. The Collaborative Adaptive Sensing of the Atmosphere (CASA) is a National Science Foundation Engineering Research Center that is conducting research on weather hazard forecasting and warning technology using low-cost radars that work at short range and adapt to evolving weather and to changing user needs (McLaughlin et al. 2009). CASA developed the Integrated Project One (IP1), which deployed a four-radar testbed in southwest Oklahoma. The four radars are spaced approximately 30 km apart and arranged in a manner to approximate equilateral triangles when connecting the radars with lines. This arrangement maximizes the regions suited for dual-Doppler velocity vector retrievals. These regions correspond to crossing azimuths from different radars that create 90 deg angles. If two radars are situated in a perfect horizontal line, then the best regions for dual-Doppler retrievals are due north and south of the midpoint between the radars.

FIGURE 5.3
The University of Oklahoma Advanced Radar Research Center's PX-1000 transportable radar.
The PX-1000 is trailer mounted, and the antenna is protected with a radome. It is quite suitable
for a long (e.g., seasonal) field campaign in remote areas.

A comparison between CASA and WSR-88D radars is given in Table 5.5.
CASA radars operate at X-band and use very low power magnetron transmitters with average power of 13 watts. The most unique aspect of the CASA
network design is the intelligent scanning based on the weather and user
needs. Measurements from a single CASA radar are heavily attenuated to the
point where signals are completely lost within a single, strong thunderstorm.
Attenuated polarimetric measurements can be corrected using Φ_{dp}-based
methods, but a signal is required to correct. The situation of total signal loss is
handled in a radar network design by having a radar on the back side of the
strong convection focus sector scans in order to adaptively fill in the weather-
caused gap in radar coverage. The CASA network design offers a solution for
gap-filling radars. The finer spatiotemporal resolution may also be required in
urban areas that respond to rainfall on the order of minutes instead of hours.

5.2 Spaceborne Radars

5.2.1 Precipitation Radar aboard TRMM

Although weather radars have been developed ever since World War II to
observe precipitation and have proven their value to the weather community

TABLE 5.5

Characteristics of CASA Radars (Middle Column) and WSR-88Ds (Right Column)

Transmitter	Magnetron	Klystron
Frequency	9.41 GHz (X-band)	2.7–3.0 GHz (S-band)
Wavelength	3.2 cm	10 cm
Peak radiated power	10 kW	500 kW
Duty cycle (max)	0.0013	0.002
Average radiated power	13 W	1000 W
Antenna size	1.2 m	8.5 m
Antenna gain	36.5 dB	45.5 dB
Radome size	2.6 m	11.9 m
Polarization	Dual linear, SHV	Dual linear, SHV
Beamwidth	1.8 deg	0.925 deg
PRF	Dual, 1.6–2.4 kHz	Single, 322–1282 Hz
Pulse width	660 ns	1600–4500 ns
Doppler range	40 km	230 km
Range increment	100 m	250 m*/1000 m
Azimuth increment	1 deg	0.5 deg*/1 deg
Scan strategy	60–360 deg adaptive PPI sector scans, 1–30 deg RHI scans	360 PPI scans, 0.5–19.5 deg elevation

* These finer resolutions are available with the NEXRAD "superresolution" upgrade.

and beyond, reliable ground-based precipitation measurements are difficult to obtain over all regions of the world, including vast oceanic and mountainous regions. The limitations of ground weather radar systems highlight the attraction of meteorological satellites to obtain seamless regional and global precipitation information for tropical rainfall studies, weather forecasting, modeling the hydrological cycle, and climate studies. The first meteorological satellite was launched in 1960, initiating a new era in space-based remote sensing of the atmosphere. A joint mission between the National Aeronautics and Space Administration (NASA) and the Japan Aerospace Exploration Agency (JAXA) launched the Tropical Rainfall Measuring Mission (TRMM) on November 27, 1997. TRMM was originally motivated by the need to understand regional and climatological rainfall patterns over previously unobserved regions in the tropics. The precipitation radar (PR) is one of the primary instruments onboard the TRMM low Earth-orbiting satellite. The PR is the first spaceborne weather radar dedicated to measuring three-dimensional structures and surface precipitation rates in tropical precipitation systems.

TRMM started as an experimental mission with an originally anticipated life-span of three to five years. The scientific community quickly realized the potential of the quasiglobal rainfall measurements, especially accumulated over longer time periods. By 2001, TRMM scientists faced an end of the mission in 2002 or 2003 due to lack of fuel. To continue the collection of high-resolution information provided by TRMM from its combination of active and passive instruments, the TRMM science teams proposed increasing the orbital altitude from 350 to 402.5 km to decrease atmospheric drag and therefore extend the lifetime of the mission. In addition to a wealth of discoveries on tropical rainfall characteristics, the TRMM products have been used to calibrate and integrate precipitation information from instruments aboard multiple polar-orbiting satellites. The real-time availability of TRMM products is used by operational weather agencies in the United States and around the world. The rainfall estimates serve a number of applications including real-time flood and landslide prediction systems (Hong et al. 2007a, 2007b; Hong and Adler 2008; Wu et al. 2012).

The PR is the first weather radar in space and was a predecessor to the second dual-frequency precipitation radar contributing to the Global Precipitation Measurement (GPM) mission. The PR is a 128-element, phased-array radar that operates at Ku-band. The PR is able to slightly adjust the Ku-band frequency to obtain 64 independent samples with a fixed PRF of 2776 Hz. The PR antenna electronically scans in the cross-track direction over ±17 deg about nadir (vertically pointing down) resulting in a swath width of 215 km. The PR has a beamwidth of 0.71 deg with a horizontal near-surface pixel resolution of 4.4 km at nadir and approximately 5 km at the edge of the scan. The range resolution of PR is 250 m in the vertical throughout the depth of storms from 20 km AGL down to the surface. TRMM travels in a non-sun-synchronous orbit from 35 deg S latitude to 35 deg N latitude, providing a revisit frequency of 11–12 hr. The primary observational goals of PR are (1) to provide detailed 3-D storm structures and (2) to obtain high-quality QPE over land as well as over ocean. The TRMM core satellite also carries a multichannel passive microwave radiometer called the TRMM Microwave Imager (TMI) and a visible and infrared scanner (VIRS), which are useful for measuring precipitation. The swath geometry of the TRMM instruments is shown in Figure 5.4.

The Precipitation Processing System (PPS) is a software infrastructure developed at NASA to process the PR, TMI, VIRS, combined instrument, and multisatellite standard products. The final build of TRMM's precipitation algorithms is referred to as the Version 7 algorithms. PPS obtains the raw data to generate the Level 1 radiance products. These are used to produce the instantaneous rainfall-related Level 2 products. Level 3 products combine data from the PR overpasses with the constellation of passive microwave instruments, geostationary satellites, and rain gauge networks on the surface. These latter products are gridded at 0.25 × 0.25 deg and are available at all grid points between 50 deg N-S latitude every 3 hr. The Level 3 products are used for climatological tropical rainfall studies and for hydrologic applications.

The quasiglobal availability of data from TRMM has led to a myriad of studies that use ground validation instruments (rain gauges, disdrometers, ground radars) to evaluate the level 2 and 3 rainfall measurements and derived products. A variety of methods have been developed to align spaceborne and ground radar data so as to compare their observations (Bolen and

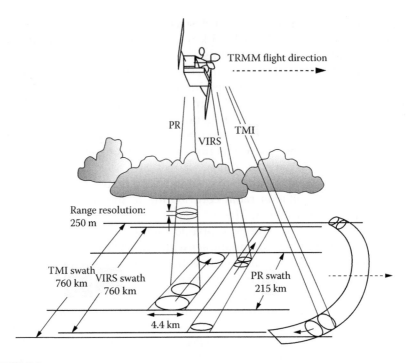

FIGURE 5.4
The TRMM satellite with its three primary instruments onboard, including the first space-borne precipitation radar, PR. TRMM was launched in 1997 with an expected mission life-time of 3–5 years. Steps have been made to conserve batteries and extend its lifetime so that it continues to operate through 2014. (Figure courtesy of NASA.)

Chandrasekar 2003). Schumacher and Houze (2000) compared the areas of echo coverage by PR and a ground radar (GR) and found that PR can capture the main rain regions but failed to detect some of the weaker echo regions. Amitai et al. (2009) conducted a comparison between the PR and GR probability distribution functions (pdfs) of the instantaneous rain rate and showed the pdfs of PR are generally shifted toward lower rain rates, indicating underestimation at the highest rain rates. In discussing differences found between TRMM PR and GR rainfall estimates in these studies, several reasons are suggested such as calibration differences, scattering differences, volume matching mismatches, errors in the attenuation correction methods, inaccurate reflectivity-to-rainfall relationships, physical properties of hydrometeors like their phase state, nonuniform beam-filling impacts, and others. Wen et al. (2011) incorporated hydrometeor classification information from a ground-based polarimetric radar to classify the PR-GR comparisons as a function of hydrometeor type. They found that PR underestimated with large diameter, wetted hydrometeors such as rain/hail mixture, wet snow, and graupel. Kirstetter et al. (2012) developed a framework for comprehensive evaluation of TRMM PR products using the MRMS products discussed in Chapter 4. The basic approach of the framework is to not only aggregate MRMS rain rates up to the PR pixel resolution, but to conduct heavy filtering and censoring of the MRMS data to maximize their quality. The first step of processing computes the spatially distributed, gridded bias that was computed in the gauge correction scheme of the MRMS algorithm. This bias was computed from hourly radar-gauge comparisons. The underlying assumption is that the radar bias does not exhibit substantial variability at the subhourly time scale. This hourly bias is then applied downscale to the instantaneous (2 min) rainfall rate fields. If the computed bias is too large (<0.1 or >10), then the correction is deemed to be too large and the pixel is discarded. The next filter screens out all pixels that have associated radar quality index (RQI) values <1. Recall from Chapter 4 that the RQI is reduced in areas where there is partial beam blockage and where the beam is sampling within and above the melting layer. The MRMS rain rates are sampled to the resolution of the PR pixel by selecting all pixels within a 2.5 km radius of the center of the PR pixel. On average, the search locates 25 MRMS pixels. If more than 5 of the pixels have missing rain rate values, then the comparison is discarded. The MRMS rain rates are then averaged using a weighting scheme that mimics the PR antenna pattern. Once the weighted mean value is computed from the MRMS reference, it is compared with the standard deviation computed from the rain rate distribution within the approximate 25 pixel neighborhood. If the weighted mean rain rate is less than its standard deviation, then the MRMS reference is considered to be nonrobust and subsequently discarded. This high rainfall variability within the PR pixel resolution will affect the retrieval, resulting in large uncertainties. Although a great number of MRMS pixels were removed in the censoring steps, the remaining dataset is accurate, has a large sample size, and is independent from PR. No processing such as statistical downscaling was performed on the PR data.

These censored MRMS datasets have been used to reveal and quantify error characteristics with TRMM PR. Kirstetter et al. (2013) compared two different versions of PR rainfall products (Versions 6 and 7). They found that the latest version improved over its predecessor by retuning the Z–R relation, which corrected PR overestimation at light rain rates (<10 mm hr^{-1}). Version 7 simultaneously corrected PR underestimation at high rain rates (>30 mm hr^{-1}) by improving the algorithm that deals with nonuniform beam-filling effects. Kirstetter et al. (2014) examined the impacts of subpixel rainfall variability on PR rainfall estimates in terms of detectability, precipitation classification (stratiform vs. convective), and quantification. These characteristics were evaluated based on the rain fraction (in %) and inhomogeneity (proxy for nonuniform beam filling) of the MRMS-observed rainfall within the PR field of view (FOV). Detection of rainfall by PR is successful if more than 70% of the FOV is filled with nonzero rain rates. They used the MRMS precipitation types to find that PR falsely detected convection in instances of low filling of the FOV and low values of NUBF. In terms of rainfall quantification, their main finding was the stratiform and convective profiling algorithms with PR seem to be lacking sufficient dynamics to deal with extreme rainfall amounts. These situations are particularly challenging due to anomalous drop size distributions (DSDs), unusually strong attenuation of the signal, and strong horizontal gradients causing large NUBF effects.

The TRMM PR products are considered the "calibrators" for the passive microwave rainfall estimation algorithms and the combined Level 3 products. Thus, errors with PR will cascade to the other rainfall products. The remote-sensing estimates are bias corrected with rain gauges, but this adjustment is performed with monthly totals and is thus not available in real time. Similar to the attention given to the radar-only product in MRMS, it is quite important to continue to improve the remote-sensing products to maximize their use in real-time applications such as hydrologic models. TRMM-based multisatellite data are being used as input into hydrological and land surface models to better understand the impacts of mass and energy fluxes between the land surface and atmosphere on time scales from days to years (Rodell et al. 2004). The TRMM Level 3 productshave been used in global flood and landslide monitoring systems (Hong et al. 2007a, 2007b; Hong and Adler 2008; Wu et al. 2012). The quasiglobal hydrologic forecasting model described in Wu et al. (2012) has been used for operational flood monitoring and prediction in a number of countries. Many of these developing countries do not have operational ground radar networks or rain gauge networks with sufficient spatiotemporal resolution to adequately resolve flood-producing rainfall. Thus, the TRMM Level 3 products, despite their uncertainties, offer the first rainfall climatologies for many countries and provide real-time rainfall estimates across national scales. Furthermore, as will be described in more detail in Chapter 8, the availability of 3 hr rainfall estimates since 1997 can be used to establish flood frequency estimates using hydrologic simulations at ungauged locations.

5.2.2 Dual-Frequency Precipitation Radar aboard NASA GPM

Success of the TRMM program has warranted an ambitious Global Precipitation Measurement (GPM) constellation mission (http://gpm.gsfc. nasa.gov) successfully launched in 2014. GPM is an extension to the TRMM mission by providing more accurate precipitation estimates using advanced instruments with an additional goal of quantitatively estimating falling snow. The GPM core observatory will be deployed in a non-sun-synchronous orbit at a 65 deg inclination and a mean altitude of 407 km. The core spacecraft will carry a dual-frequency, phased-array precipitation radar (DPR) to provide measurements of 3-D precipitation structures and microphysical properties in precipitating clouds. The DPR operates at Ka-band (35.5 GHz) and Ku-band (13.6 GHz) frequencies. They provide 3-D measurements of precipitation structures at 5 km footprints (at nadir) over a cross-track swath width of 120 km for the Ka-band radar and 245 km for Ku-band. The beamwidth for both frequencies is the same as TRMM PR at 0.71 deg. The pulse repetition frequency (PRF) for both radars is 4100–4400 Hz. The variable PRF method is expected to improve the lower sensitivity of the retrievals, which is particularly important for snowfall retrievals. The pulse width is 1.667 μs for Ku-band and 1.667/3.234 μs at Ka-band. These pulse widths result in range (vertical) resolutions of 250 m and 250/500 m at Ku- and Ka-bands, respectively.

The GPM core spacecraft, which will carry the DPR, and an advanced 13 channel microwave radiometer will provide precipitation estimates between 65 deg N-S latitude. A goal of the GPM mission is to improve passive microwave-based retrievals over land, especially for mid- and high-latitude regions. The GPM microwave imager (GMI) aboard the core spacecraft is similar to TMI but has higher frequency channels up to 183 GHz. It also has a larger antenna, which improves spatial resolution. The primary advancement with the DPR will be the use of a dual frequency ratio (DFR), which is merely the difference in Z at Ku- and Ka-bands. The signals attenuate in a precipitation medium at different rates. It turns out this difference in attenuation is related to characteristics of the DSD, in particular the median drop diameter. This variable will be useful for DSD retrievals, precipitation rate estimation, and rain-snow segregation. Moreover, it may be possible to compare these spaceborne measurements to NEXRAD polarimetric measurements (i.e., Z_h and Z_{DR}). The independent measurements from space may be used to calibrate NEXRAD polarimetric variables just as they were used to identify miscalibrated NEXRAD radars with comparisons of space-based and ground-based Z (Anagnostou et al. 2001; Bolen and Chandrasekar 2003).

5.3 Phased-Array Radar

5.3.1 Design Aspects and Product Resolution

The NEXRAD network presently comprises dual-polarization radars with conventional pedestals and antenna designs. The network has already exceeded its 20-year engineering design life-span. Components of the WSR-88D that require regular maintenance are the moving parts, primarily the pedestal and rotary joints. Similar to the background of conventional weather radar being rooted in the military, phased-array radar (PAR) technologies have been around for several decades. They were developed primarily for the purpose of detecting very fast moving targets coming from multiple directions simultaneously, such as aircraft and missiles. PAR is a potential successor technology for NEXRAD. It offers the following advantages over conventional weather radars: (1) updates of volume scans at intervals on the order of seconds instead of minutes, and (2) ability to focus on different sectors as well as multimission use to track airplanes simultaneously (Zrnić et al. 2007). These advances will lead to improvements in weather monitoring for rapidly changing phenomena such as supercell storms, downbursts, and wind events. Moreover, depending on the design of the array, it may be possible to electronically steer the beams in all directions, thus negating the need to rotate the array with a pedestal.

The main difference between PAR and conventional radars that use a mechanically rotating parabolic antenna lies in the way the beams are directed or steered. The PAR beam is formed and transmitted electronically by controlling the phase and pulsing of the individual transmit-receive elements. Consider an example where a flat panel containing the array of elements is facing due north. A beam can be steered to the northeast direction if the elements on the west side of the array transmit first, followed by elements in the middle and then to the east. This built-in west-to-east delay causes superposition of the electromagnetic waves so that the net effect causes the beam to be effectively steered to the northeast. The same concept can be applied to any direction 45 deg offset from the panel's pointing axis (due north from the example just provided). Moreover, some elements can be dedicated to a given regime, or storm, while another group focuses on a different sector. This **beam agility** is what enables very quick updates and multimission functionality.

To develop, test, and demonstrate the advantages of phased-array technology for operational surveillance, a National Weather Radar Testbed (NWRT) was established in Norman, Oklahoma (Figure 5.5). The NWRT consists of a converted U.S. Navy SPY-1A phased-array antenna, a modified WSR-88D transmitter, and a custom radar processor (Zrnić et al. 2007). The antenna consists of 4352 elements that steer the beam. It is a single panel providing 90 deg of coverage in the azimuthal direction, so it must be rotated with a pedestal to complete a full volume scan. Elevation scans are accomplished using electronic scanning by lagging the transmitted pulses from bottom to top of the panel. Characteristics of the NWRT are summarized in Table 5.6. This phased-array radar, which supports oversampling in range by a factor of 10, can record time-series data and is controlled remotely.

5.3.2 Dual Polarization

The ability to obtain dual-polarization measurements generally requires simultaneous transmission and reception of the signal or at least a very fast switch to produce matched beam patterns at horizontal and vertical polarizations. This is a requirement that poses significant challenges to phased-array radars. The measurement errors in polarimetric variables are

FIGURE 5.5
Phased-array radar antenna of the National Weather Radar Testbed (NWRT) comprising 4352 components used to electronically steer the beam. (Figure courtesy of Zrnić et al. 2007).

TABLE 5.6

Characteristics of the National Weather Radar Testbed Phased-Array Radar

Transmitting antenna diameter	Approximately 3.66 m (≈ circular aperture)
Wavelength	9.38 cm (S-band)
Transmitting beamwidth	Approximately 1.5 deg (up to 2.1 deg at 45 deg from beam center)
Receiving beamwidth	Approximately 1.66 deg (larger than transmitting beamwidth to reduce sidelobes)
Transmitter power and pulse width	About 750 kW peak and 1.57 μs or 4.71 μs
Sensitivity	Reflectivity of 5.9 dBZ at 50 km produces a SNR = 0 dB

Source: From Zrnić et al. (2007).

already significant with conventional parabolic antennas, especially for Z_{DR}. If a planar phased-array radar is used, such as the NWRT, beams that are displaced off the pointing angle (called broadside) have larger beamwidths, less sensitivity, and a limited basis for obtaining polarized measurements from nonorthogonal waves. Zhang et al. (2011) proposed a cylindrical polarimetric

FIGURE 5.6
Prototype of the cylindrical polarimetric phased-array radar. The cylindrical design for phased-array radar provides a solution for maintaining resolution while collecting dual-polarization radar moments.

phased-array radar (CPPAR) design that avoids many of the issues encountered with the planar design. The cylindrical design maintains its basis for orthogonal polarization and has the same beamwidth and sensitivity at all azimuths. This radar design was recently built and demonstrated on a mobile platform (Figure 5.6).

5.3.3 Impact on Hydrology

The greatest advantage in phased-array radar measurements for hydrologic applications is the ability to provide rainfall estimates at a frequency less than 1 min. Note it takes a WSR-88D radar approximately 5 min to complete an entire volume scan. The MRMS algorithm described in Chapter 4 yields rainfall rate products at a 2 min resolution. But the highest frequency can be achievable only for regions that are covered by two or more radars, and the two radars must be operating out of sync. This means they'll be providing independent radar measurements over the overlapping region, thus yielding rainfall rate estimates at frequencies higher than 5 min. A network of phased-array radars would be able to produce rainfall rates at frequencies higher than 1 min. This may have the greatest application in small, urban basins that respond very quickly to rainfall.

Figure 5.7 shows a time series of rainfall estimates computed over a medium-sized 813 km² catchment using phased-array radar measurements, but resampled at different time intervals. We can see the impact of temporal sampling of the radar-estimated rainfall rates on basinwide rainfall. At this basin scale, the 15 min rainfall rates appear very similar to the 5 min ones. The peak amounts are approximately double the rates with the hourly sampling. This result indicates that sampling subhourly is important for this basin scale. The next step in such an exercise would be to input the rainfall rates sampled at different frequencies into a hydrologic model and compare results to observed streamflow. In any case, the convergence of the time series at temporal resolutions finer than 15 min suggests that 1 min sampling over a medium-sized basin would not necessarily have an impact on hydrologic simulations. The impact of space-time resolution of precipitation forcing on hydrologic response depends on basin scale, among other factors such as relief, soil types and depth, and land cover.

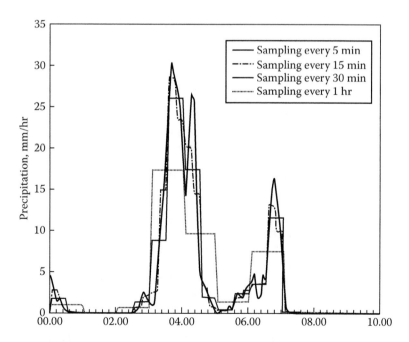

FIGURE 5.7
Impact of temporal frequency of rainfall rates estimated by a phased-array radar on basinwide rainfall.

Problem Sets

QUALITATIVE QUESTIONS

1. Describe the advantages and disadvantages of weather radars in measuring precipitation with different frequencies: X-band (3 cm), C-band (5 cm), and S-band (10 cm). (Hint: Consider the difference in radar/antenna size, transmitter power, radar range, Earth curvature effect, radar resolution, Rayleigh/Mie scattering, precipitation attenuation, etc.)

2. Briefly describe the advantages and disadvantages of small mobile radars compared with fixed-site S-band weather radars.

3. Briefly describe the advantages of GPM compared with TRMM.

4. What are the main characteristics of phased-array radar?

5. What are the challenges and opportunities of phased-array polarimetric radars compared with conventional radars?

References

Amitai, E., X. Llort, and D. Sempere-Torres. 2009. Comparison of TRMM radar rainfall estimates with NOAA next-generation QPE. *Journal of the Meteorological Society of Japan* 87A: 109–118.

Anagnostou, E. N., C. A. Morales, and T. Dinku. 2001. The use of TRMM precipitation radar observations in determining ground radar calibration biases. *Journal of Atmospheric and Oceanic Technology* 18: 616–628.

Biggerstaff, M. I., L. J. Wicker, J. Guynes, C. Ziegler, J. M. Straka, E. N. Rasmussen, A. D. IV, L. D. Carey, J. L. Schroeder, and C. Weiss. 2005. The Shared Mobile Atmospheric Research and Teaching Radar: A collaboration to enhance research and teaching. *Bulletin of the American Meteorological Society* 86 (9): 1263–1274.

Bolen, S. M., and V. Chandrasekar. 2003. Methodology for aligning and comparing spaceborne radar and ground-based radar observations. *Journal of Atmospheric and Oceanic Technology* 20: 647–659.

Bousquet, O., A. Berne, J. Delanoe, Y. Dufournet, J. J. Gourley, J. Van-Baelen, C. Augros, L. Besson, B. Boudevillain, O. Caumont, E. Defer, J. Grazioli, D. J. Jorgensen, P.-E. Kirstetter, J.-F. Ribaud, J. Beck, G. Delrieu, V. Ducrocq, D. Scipion, A. Schwarzenboeck, and J. Zwiebel. 2014. Multiple-frequency radar observations collected in southern France during the field phase of the hydrological cycle in the Mediterranean experiment (HyMeX). *Bulletin of the American Meteorological Society* (In press).

Cheong, B. L., R. D. Palmer, M. Yeary, T.-Y. Yu, and Y. Zhang. 2011. Design, fabrication and test of a TWT transportable polarimetric X-band radar. *Proceedings 91st Annual Meeting*, American Meteorological Society, Seattle, Washington, January.

Ducrocq, V., et al. 2014. HyMeX-SOP1, the field campaign dedicated to heavy precipitation and flash flooding in the northwestern Mediterranean. *Bulletin of the American Meteorological Society* 95(7): 1083–1100.

Gourley, J. J., D. P. Jorgensen, S. Y. Matrosov, and Z. L. Flamig. 2009. Evaluation of incremental improvements to quantitative precipitation estimates in complex terrain. *Journal of Hydrometeorology* 10: 1507–1520.

Hong, Y., R. F. Adler, F. Hossain, S. Curtis, and G. J. Huffman. 2007a. A first approach to global runoff simulation using satellite rainfall estimation. *Water Resources Research* 43 (8): W08502.

Hong, Y., R. Adler, and G. Huffman. 2007b. An experimental global prediction system for rainfall-triggered landslides using satellite remote sensing and geospatial datasets. *IEEE Transactions on Geoscience and Remote Sensing* 45 (6): 1671–1680.

Hong, Y., and R. F. Adler. 2008. Predicting global landslide spatiotemporal distribution: Integrating landslide susceptibility zoning techniques and real-time satellite rainfall estimates. Special Issue of *International Journal of Sediment Research* 23 (3): 249–257.

Isom, B., R. Palmer, R. Kelley, J. Meier, D. Bodine, M. Yeary, B. L. Cheong, Y. Zhang, T.-Y. You, and M. I. Biggerstaff. 2013. The Atmospheric Imaging Radar: Simultaneous volumetric observations using a phased array weather radar. *Journal of Atmospheric and Oceanic Technology* 30: 655–675.

Kirstetter, P.-E., Y. Hong, J. J. Gourley, S. Chen, Z. L. Flamig, J. Zhang, M. Schwaller, W. Petersen, and E. Amitai. 2012. Toward a framework for systematic error modeling of spaceborne precipitation radar with NOAA/NSSL ground radar-based National Mosaic QPE. *Journal of Hydrometeorology* 13: 1285–1300.

Kirstetter, P.-E., Y. Hong, J. J. Gourley, M. Schwaller, W. Petersen, and J. Zhang. 2013. Comparison of TRMM 2A25 products, version 6 and version 7, with NOAA/NSSL ground radar-based National Mosaic QPE. *Journal of Hydrometeorology* 14: 661–669.

Kirstetter, P.-E., Y. Hong, J. J. Gourley, M. Schwaller, W. Petersen, and Q. Cao. 2014. Impact of sub-pixel rainfall variability on spaceborne precipitation estimation: evaluating the TRMM 2A25 product. *Quarterly Journal of the Royal Meteorological Society* (Accepted).

McLaughlin, D., et al. 2009. Short-wavelength technology and the potential for distributed networks of small radar systems. *Bulletin of the American Meteorological Society* 90: 1797–1817.

Palmer, R., M. Biggerstaff, P. Chilson, G. Zhang, M. Yeary, J. Crain, T.-Y. Yu, Y. Zhang, K. Droegemeier, Y. Hong, A. Ryzhkov, T. Schuur, and S. Torres. 2009. Weather radar education at the University of Oklahoma: An integrated interdisciplinary approach. *Bulletin of the American Meteorological Society* 90: 1277–1282.

Rodell, M., et al. 2004. The global land data assimilation system. *Bulletin of the American Meteorological Society* 85: 381–394.

Schumacher, C., and R. A. Houze, Jr. 2000. Comparison of radar data from the TRMM satellite and Kwajalein oceanic validation site. *Journal of Applied Meteorology* 39: 2151–2164.

Wen, Y., Y. Hong, G. Zhang, T. J. Schuur, J. J. Gourley, Z. L. Flamig, K. R. Morris, and Q. Cao. 2011. Cross validation of spaceborne radar and ground polarimetric radar aided by polarimetric echo classification of hydrometeor types. *Journal of Applied Meteorology and Climatology* 50: 1389–1402.

Wu, H., R. F. Adler, Y. Hong, Y. Tian, and F. Policelli. 2012. Evaluation of global flood detection using satellite-based rainfall and a hydrologic model. *Journal of Hydrometeorology* 13 (4): 1268–1284. ·

Zhang, G., R. J. Doviak, D. S. Zrnić, R. Palmer, L. Lei, Y. Al-Rashid. 2011. Polarimetric phased-array radar for weather measurement: A planar or cylindrical configuration? *Journal of Atmospheric and Oceanic Technology* 28: 63–73. doi: http://dx.doi.org/10.1175/2010JTECHA1470.1.

Zrnić, D. S., J. F. Kimpel, D. E. Forsyth, A. Shapiro, G. Crain, R. Ferek, J. Heimmer, W. Benner, T. J. McNellis, and R. J. Vogt. 2007. Agile beam phased array radar for weather observations. *Bulletin of the American Meteorological Society* 88 (11): 1753–1766.

6

Radar Technologies for Observing
the Water Cycle

6.1 The Hydrologic Cycle

The **hydrologic cycle** describes the location, movement, and flux of water through and across the Earth's surface, oceans, and overlying atmosphere. The major fluxes include precipitation, evapotranspiration, river discharge, infiltration, and groundwater flows. It is important to track and quantify the partitioning of water throughout its course in the hydrologic cycle for several reasons. Freshwater is the component of the water on Earth that sustains life for terrestrial plants and animals. This precious resource comprises only 3% of the total water available on Earth, and most of that small percentage of water is locked in ice caps and glaciers (~69%) or stored as groundwater (~30%). That leaves ~0.3% of the Earth's total freshwater stored on the surface in lakes, rivers, and swamps. Monitoring this resource is vital to sustaining life, especially in regions that are prone to extended drought and subsequent water shortages. Second, knowledge of the water partitioning within the hydrologic cycle gives a pulse or fingerprint of the climate state of the planet. A cold climate state is associated with phase changes from liquid to ice and yields large polar ice caps, glaciers, and the ice ages that have been documented in the past. Warmer climate states have less ice and more oceanic water. By detecting these small changes within the hydrologic cycle, we can determine the trajectory of the Earth's changing climate system. This chapter introduces remote-sensing solutions to monitoring various components of the hydrologic cycle.

The concept of a watershed is introduced for computing a closed water balance. A watershed is defined as a bounded geographic domain, determined by topography. The generalized water balance equation for a watershed is given as

$$\Delta S = P - Q - ET \qquad (6.1)$$

where P is precipitation input, Q is river discharge, and ET is evapotranspiration. ΔS is the storage term and can represent multiple water storages in various compartments including the soils, underlying aquifer, snowpack, lakes, and swamps. The storage term is more difficult to measure and quantify, so it is often treated as the residual term in Equation (6.1). This simple water balance becomes much more complicated for watersheds that have significant anthropogenic impacts. For instance, groundwater pumping can remove a great deal of water stored in the underlying aquifer and can transport it out of the basin or can store it on the surface to become a large source of *ET* or *Q*. River water can also be stored or redirected for municipal or agricultural uses such as irrigating crops.

Precipitation is measured using in situ instruments such as rain gauges or with radar as has been discussed in prior chapters. River discharge is conventionally measured using an in situ floating device that provides the depth (or stage) of the river. This stage is converted to a discharge value (volume per unit time) using a rating curve. The rating curve is established for each gauge by visiting the site and taking manual measurements during different times of the year of the stream velocity, stage height, and stream cross-section (or bathymetry). A rating curve for the Blue River Basin in south-central Oklahoma is shown in Figure 6.1. The circles correspond to the individual measurements. We can see at lower flows that four different modes indicate variable relationships between stage height and discharge. This variability exists due to changes in the riverbed from sediment movement, aquatic vegetation, and braiding of the stream. These factors that lead to variability in the relationship become negligible for high, flooding flows. After a regression curve is established and regularly updated, it is used to convert the automatic observations of stage to a more useful value of discharge. Later in this chapter, the concept of river discharge estimation using radar will be introduced.

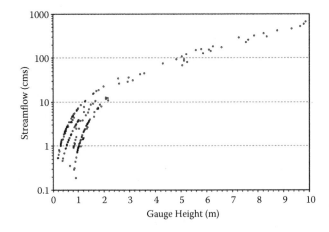

FIGURE 6.1

Manually measured rating curve that relates automatic measurements of stage height (in m) to volumetric flow rate, or streamflow (m³ s⁻¹, abbreviated as cms) for the Blue River basin in south central Oklahoma (USGS #07332500).

ET is the flux of water vapor from the top-layer soils and from photosynthetically active vegetation. It is more difficult to directly measure and is commonly estimated using a number of methods that typically depend on meteorological observations such as temperature, wind speed, relative humidity, and solar radiation. It also varies with the type of crop, stomata resistance, and the degree of water that is available to plants, which can be very high for irrigated crops. The concept of potential ET (PET) is commonly used in hydrologic modeling and water balance studies. PET is the amount of ET that would occur if the water supply were unrestricted. It tends to be greatest during hot, sunny, and windy times of the year. PET is easier to estimate than ET, so it is common practice to estimate PET and then reduce it based on the water availability, which can be either modeled or estimated from soil moisture measurements.

A simple monthly water balance using Equation (6.1) was computed for the Blue River basin in Figure 6.2. The monthly basin-averaged precipitation time series shows the peak in the late spring followed by a secondary, smaller peak in the autumn months. The estimated loss due to *ET* approximately mirrors the *P* trend throughout the year but is of lower magnitude. The mean discharge, which has been normalized by the watershed area, is substantially smaller than the inputs from *P* and outputs from *ET*. The *Q* values lag the input *P* during the autumn months. The Δ*S* residual correspondingly reaches maximum values during the early autumn months. This indicates that water

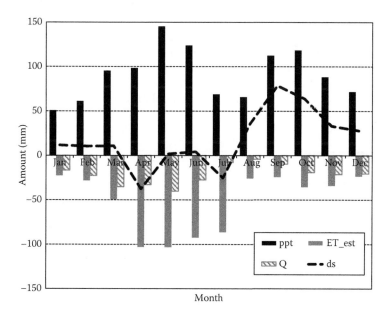

FIGURE 6.2
Monthly water balance on the Blue River basin in south central Oklahoma using in situ measurements of variables in Equation (6.1), except that the Δ*S* storage term was computed as a residual.

is being stored following the hot, dry summer months. It turns out the Blue River has a karstic geologic formation with a high degree of connectivity between surface water and groundwater. The basin goes into a storage mode following the summer so that incoming water recharges the aquifer. As we can see, the simple water balance equation has yielded insights into the hydrologic behavior of the basin. All of the observations used up to conduct this analysis were from in situ measurements (i.e., meteorological observations, rain gauges, stream gauge). Note that the observations were climatological values based on decades of data. For such a large time scale, it is feasible to use in situ data because a point measurement is more representative of a large spatial scale. However, as we come down in scale to study or forecast the hydrologic response to a given storm's precipitation inputs, it becomes necessary to resolve the spatiotemporal patterns of the hydrologic variables. This is called **distributed hydrology** and readily accommodates hydrologic observations from radars.

Radar precipitation and microphysical studies have been accomplished using ground, airborne, and space-based radars with wavelengths ranging from W- to S-band. Radar-based quantitative precipitation estimates (QPEs) are used ubiquitously for hydrologic studies and as inputs for operational systems to monitor and predict flash flooding. In this chapter, we discuss radar technologies that can be used to monitor additional stores and fluxes of water in the hydrologic cycle. This includes streamflow, surface water depth and spatial extent, top-layer soil moisture, root-zone soil moisture, and depth to the groundwater table. One component of the hydrologic cycle whose measurement remains elusive to radars is *ET*. In general, increasingly longer radar wavelengths are used to detect water from the top of the cloud all the way down to the water table (see Table 1.1). In several cases, the radars must be pointed downward toward the Earth. These measurements are made more readily from airborne and spaceborne platforms.

6.2 Surface Water

6.2.1 Streamflow Radar

In recent years, new approaches have evolved within the hydrologic community for streamflow estimation. More economical methods of measuring stream discharge through remote sensing include acoustic Doppler profilers (Simpson and Oltmann 1993; Yorke and Oberg 2002). These instruments are typically deployed in the bottom of a stream and provide more accurate depictions of stream velocity with depth. There have also been recent advances in noncontact methods of stream gauging using radars and particle image velocimetry (Costa et al. 2006; Creutin et al. 2003). Large-scale particle image velocimetry (LS-PIV) relies on the use of optical instruments (e.g., cameras) that are mounted on bridges looking down at the water surface. Software is used to essentially track the motion of bubbles on the surface and compute their velocity. The LS-PIV method is inexpensive but requires a known river cross-section including measured stage height to accurately estimate streamflow.

Costa et al. (2006) presented a noncontact radar solution for measuring stream surface velocities by comparing results from a cable-mounted 9.36 GHz pulsed Doppler radar, 350 MHz monostatic UHF, and a continuous wave 24 GHz microwave system to in situ streamflow data. The surface velocities were converted to mean velocity (with depth) using profiles measured by current meters and an acoustic Doppler profiler. The stream cross-section is also needed to compute discharge. For this, they used a collocated, nadir-pointing 100 MHz ground-penetrating radar (GPR). In low-conductivity water, they were able to measure the river cross-section within 1%–5% by moving the GPR on a cable across the river. This information (which can be obtained infrequently) was combined with the near-continuous stream velocity measurements (and stage heights) to yield discharge measurements within 5% of the in situ gauges. The advantages they pointed out in their exploratory study were lower costs, safer measurements (noncontact), higher frequency, and potentially more accurate measurements using remote-sensing methods. The constraints discussed were issues in penetrating the water column to the bottom of the channel with the GPR in high-conductivity river water (more likely common in sediment-rich floods) and the requirement of a cableway for the GPR to traverse the river.

A conceptual stream radar shown in Figure 6.3 builds upon the principles of noncontact radar measurements discussed above, but taking all required measurements (i.e., river cross-section, stage height, and mean velocity) using

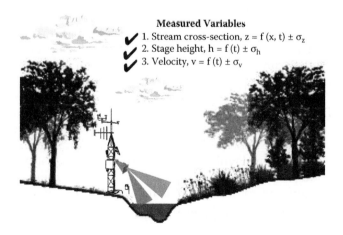

Measured Variables
1. Stream cross-section, $z = f(x, t) \pm \sigma_z$
2. Stage height, $h = f(t) \pm \sigma_h$
3. Velocity, $v = f(t) \pm \sigma_v$

FIGURE 6.3
Prototype depiction of a dual-frequency, scanning stream radar capable of simultaneously measuring near-surface stream velocity, stage height, and the riverbed cross-section.

a single radar system. A dual-frequency (UHF, Ku-band) scanning Doppler radar will be mounted on a tower near a stream. The radar will use the UHF channel when pointing near nadir and scanning perpendicular to the stream to obtain the channel cross-section (or bathymetry). These measurements will need to penetrate the water column to the channel bed and thus may have limitations for use in deep, sediment-rich rivers. The cross-section data would be needed only on an occasional basis (e.g., weekly). Then, the radar will use the Ku-band channel and the same cross-sectional scanning to retrieve the stage height. This higher frequency will be unable to penetrate the water column, so the stage height is computed as the difference between the range to the top of the water column (via Ku band) and the bottom of the channel (via UHF). Next, the radar will scan up the stream rather than across it using the Ku-band channel. In this quasicontinuous Doppler data collection mode, the radar can collect stream velocity measurements. At this point, all variables have been measured to estimate streamflow without the need for a manually measured rating curve.

Data can be logged on-site or transmitted via radio, cell, or satellite communications. The power requirements are not significant; thus the supply can be accomplished with a solar panel with battery storage nearby. Goals and challenges of the stream radar concept involve low cost, lightweight design, and accuracy to within 10% of in situ measurements on small streams. It is possible that this remote-sensing solution to streamflow estimation could even become an instrument aboard an airborne platform, such as an unmanned aerial vehicle. This system would enable the measurement of streamflow at several points along the stream, providing unprecedented measurements in the along-stream direction.

6.2.2 Surface Water Altimetry

Radars provide a precise range to a target in the field of view. When they are pointing in the nadir direction (straight down) over oceans and terrestrial water bodies, they can provide detailed information of oceanic topography and the variations of the water surface heights in lakes, reservoirs, swamps, and rivers. This is the basis of surface water altimetry. The oceanic observations will be able to resolve circulations that are important in weather forecasting, navigation, and management of fisheries. For example, the Gulf Stream is a very well known oceanic circulation, which was incidentally discovered by Benjamin Franklin during a transatlantic ship voyage, that impacts rainfall patterns on the eastern U.S. seaboard. These oceanic circulations play significant roles in global carbon and heat exchanges with the atmosphere; thus, monitoring them will lead to insights on global climate change. Observations over land and coastal areas will impact the hydrologic community by quantifying the spatial and temporal variability of surface freshwater storages. By obtaining the slope of the water surface in a large river, it is also possible to estimate the river's volumetric discharge.

NASA and the French Centre National d'Etudes Spatiales (CNES) will jointly launch the Surface Water Ocean Topography (SWOT) mission in 2020. SWOT will carry a Ka-band radar interferometer (KaRIN). The principle of **interferometry** is employed by splitting and transmitting a radar pulse to two antennas that are separated by a known distance (e.g., 10 m in the case of SWOT). The signals are directed to the Earth's surface, where they are reflected back up and received by the antenna on the opposite side of the boom. This concept is illustrated in Figure 6.4. Very small phase differences are extracted from the received signals that are due to differences in the index of refraction or the path length itself. The radar pulses traverse very similar atmospheric conditions, so the differences are attributed to the path length, which gives the surface water height gradient. KaRIN will provide a surface water height precision of approximately 1 cm for pixels with resolutions of 2 × 10 m (far swath) to 60 m (near swath). The total swath width will be 120 km, and the revisit frequency will be 22 days.

SWOT is currently planned to be a three-year mission. Its orbital characteristics combined with the KaRIN technical specifications will be able to characterize oceanic circulations down to a spatial resolution of

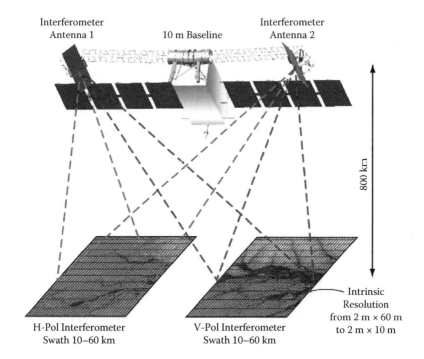

FIGURE 6.4
Retrieval of surface water elevation using the principle of interferometry. The illustration shown is the Ka-band Radar Interferometer (KaRIN) that will be the core for the Surface Water Ocean Topography (SWOT) mission. (Figure courtesy NASA Jet Propulsion Laboratory.)

15 km and larger. Current altimeter missions have a capability of resolving these circulations greater than 200 km, so SWOT will fill in a significant observational gap. Concerning terrestrial water bodies, SWOT will provide global water storage changes for lakes, reservoirs, and swamps greater than 250 m^2 in surface area. Streamflow can be estimated with the water surface gradient measurements for rivers that are at least 100 m wide and possibly 50 m. Although the revisit frequency of SWOT is only 22 days, it is possible that it will observe water elevations with floods from coastal storm surge events and inundation from flooded rivers. The primary objective of the SWOT mission is to create the first survey of Earth's surface water.

6.2.3 Synthetic Aperture Radar

Recall from Equation (1.6) that the range resolution of a pulsed radar system is determined by the pulse width. The azimuthal resolution is determined by the beamwidth, which depends on the physical size of the antenna. The pulse width can be compressed by increasing the bandwidth of the transmitting signal, which is called pulse compression, but size and weight limitations are on antennas for many platforms. For a conventional, real aperture radar (RAR), larger antennas are needed to accomplish high resolution in the azimuthal direction, but it may be impractical to mount it on an aircraft or spacecraft. A synthetic aperture radar (SAR) overcomes the antenna size limitation to obtaining high azimuthal resolution by moving it quickly and sighting perpendicular to the flight track. The concept of SAR is quite similar to a phased-array radar (PAR) that has many elements positioned on a cylindrical or planar panel. The SAR simulates a phased array by essentially moving a small antenna quickly while transmitting several pulses and assuming that the targets are standing still or have a negligible velocity compared with the antennas'. This technique has the same end result as having a large array of elements that transmit with a temporal offset; this is how the beam is electronically steered. In the case of SAR, the phase offset is achieved by physically moving the radar spatially rather than directly rearranging the phase of multiple dipole elements in the case of PAR. The end result is a synthetically built large antenna that yields high azimuthal resolution.

SAR is used for mapping features on the Earth's surface similar to objectives that can be accomplished with optical instruments. The largest advantage of SAR is the ability to collect data at night and during cloudy conditions. Benefiting from interferometry and polarization applied to SAR, (InSAR and PolSAR, respectively), it is most useful during crises such as oil spills, earthquakes, landslides, volcanoes and floods (Tralli et al. 2005). TerraSAR-X is a SAR satellite mission carried out between the German Aerospace Center (DLR) and a private space agency called EADS Astrium. The TerraSAR-X is an X-band polarimetric SAR that was launched in 2007. It flies in a sun-synchronous orbit at an altitude of 514 km. Different from optical sensors, the resolution of SAR is rarely correlated to the height of the platform because of the synthetic technique and pulse compression. Generally, the higher the carrier, the longer the synthetic path (equivalent antenna size). So even for an airborne SAR, the resolution can be as low as 6 m if the antenna size is 12 m.

The scene sizes vary from 5 × 10 km at the highest resolution up to 150 × 100 km. It has a revisit frequency of 11 days. SAR data from TerraSAR-X and platforms that operate at other wavelengths (e.g., C- and L-bands) have been used for a number of hydrologic studies including monitoring flood dynamics in urban areas, calibrating hydraulic and flood inundation models, and real-time flood management (Schumann et al. 2011; Mason et al. 2009; Stephens et al. 2011; Matgen et al. 2007; Di Baldassarre et al. 2009).

6.3 Subsurface Water

As we saw with the water budget study on the Blue River basin, water storage and later release by the karstic aquifer played a significant role in the monthly climatology of streamflow at the surface. Note that these conclusions are based on inferences from the residual ΔS term in the water balance equation (6.1) combined with knowledge of the underling geology of the basin. The capability to detect the depth to the water table and monitor its evolution over time will greatly advance hydrologic understanding of surface water and groundwater interactions. These remote-sensing measurements will lead to better management of groundwater resources and will also improve surface water monitoring and forecasts.

Top-layer soil moisture and root-zone soil moisture play significant roles in ecology, agriculture, weather forecasting, and flood-forecasting applications. As with retrievals from other types of radars, the observations must be calibrated or at least verified using in situ measurements. However, unlike other hydrological components such as precipitation, gauge-based observations for groundwater and soil moisture are sparser. The depth to the water table can be determined using a monitoring well. These are expensive to drill, and like any in situ instruments they have limitations with their spatial representations. In situ soil moisture sensors are more common than groundwater monitoring wells. They can provide data at multiple depths from 5 cm down to 75 cm below the surface. Nadir-pointing radars from airborne and spaceborne platforms offer the potential to map the spatial distribution of top-layer and root-zone soil moisture. Ground-penetrating radars are able to take deeper soil moisture measurements, as well as locate groundwater tables. These new radar technologies will provide unique observations on the spatiotemporal behavior of subsurface water, leading to new theories and better model forecasts.

6.3.1 L-Band Radar

L-band microwave remote sensing uses low frequencies to measure soil moisture in the top 0–5 cm of the surface (Colliander, et al., 2012). Testing has been done using airborne passive, active, L-, and S-band (PALS) sensors. NASA is launching the **Soil Moisture Active and Passive (SMAP)** mission in 2014, which will carry an L-band SAR as well as a passive microwave radiometer. Compared with L-band radiometer measurements, soil moisture measurements derived from L-band radar have high spatial resolution but moderate soil moisture accuracy. L-band radar is more sensitive to surface characteristics such as surface roughness, topographic features, and vegetation canopy than passive systems (Hong et al. 2012).

SMAP is one of the four first-tier missions recommended by the National Research Council's Earth Science Decadal Survey Report (Hong et al. 2012). SMAP will provide surface layer (~5 cm) soil moisture measurements, freeze/thaw states, and soil moisture at the root zone (which is simulated with a land surface model by assimilating surface soil moisture). The primary science objectives of the SMAP mission are estimating water, carbon, and energy fluxes at the land surface, improving weather and climate forecasts, and improving drought and flood monitoring capabilities (Yuen 2012).

The SMAP L-band (1.26 GHz) SAR will have transmit and receive at horizontal polarization (HH), vertical (VV), and transmit at horizontal and receive at vertical polarization (HV). The passive radiometer, also at L-band (1.4 GHz), will have H, V, U polarization. They share a 6 m diameter deployable mesh antenna with conical scanning at 13 rpm at constant incidence angle of 40 deg. The orbit will be sun-synchronous at 685 km altitude, yielding a 1000 km-wide swath. The scientific goal for integrating the data from the two instruments will result in an accuracy of 0.04 m^3 m^{-3} for volumetric water content at a spatial resolution of 10 km and temporal frequency of 2–3 days for global mapping of soil moisture (Entekhabi et al. 2010). The principles governing PALS and SMAP are similar: increased backscatter indicates higher soil moisture. This is caused by the relationship between water and soil that is discussed in more depth in Section 6.3.3 on ground-penetrating radars (Bolten et al. 2003).

6.3.2 C-Band Radar

In addition to L-band radars, C-band is another radar frequency used to estimate soil moisture in the top few centimeters of the soil. The **Advanced SCATterometer (ASCAT)** is a real-aperature C-band radar (5.255 GHz) with two vertically polarized antennas onboard the Meteorological Operation (MetOp) satellite operated by the European Organization for the Exploitation of Meteorological Satellites (EUMETSAT). The satellite's orbit at a mean altitude of 817 km yields a swath width of 550 km and achieves global coverage every 1.5 days. The primary products are wind speed and direction over the oceans, polar ice, and active storm data at 25 and 50 km spatial resolutions. Relative soil moisture (or degree of saturation) is a derived product using the Vienna University of Technology (TUWEIN) time series–based change detection algorithm, developed by Wagner et al. (1999). The algorithm applies an exponential filter to estimate the average value of the soil moisture profile using the surface soil moisture product time series. This approach assumes a linear relationship between soil moisture and the backscatter in decibel space. Wagner's exponential filter is relatively simplistic, but it is an effective method that relies on the analytical solution of a differential equation. It reliably retrieves profile soil moisture values from surface values, based on using in situ observations and modeled data. The exponential filter can solve for both the surface soil moisture (SSM) product and the root-zone soil moisture (RZSM) product. A simple version of Wagner's method to determine the root-zone soil moisture product is provided below (Brocca et al. 2011):

$$RZSM_n = RZSM_{n-1} + K_n \left[SSM(t_n) - RZSM_{n-1} \right] \qquad (6.2)$$

where the gain K_n, which varies from 0–1, at time t_n is

$$K_n = \frac{K_{n-1}}{K_{n-1} + e^{-\frac{(t_n - t_{n-1})}{T}}} \qquad (6.3)$$

where T is a characteristic time scale of soil moisture variations. To initialize the exponential filter, K_0 is set to 1 and $RZSM_0$ is set to $SSM(t_0)$. From there, the SSM and $RZSM$ can be determined.

6.3.3 Ground-Penetrating Radar

Ground-penetrating radars (GPRs) are the most common noninvasive methods to penetrate the subsurface for determining soil moisture and locating groundwater tables (Doolittle et al. 2006). Modern GPRs are small, relatively lightweight, and portable; and in most cases, only one or two people are needed to operate a GPR. Much of the research pertaining to GPR has occurred within the past two decades and is evolving rapidly as research continues. GPRs operate by sending a pulsed electromagnetic wave into the subsurface that reflects off layers or objects with a high dielectric permittivity. With this technology, GPRs can penetrate down to 30 m in certain soils under conducive conditions. Typical frequencies range from 50 to 1200 MHz, though the range can be as low as 10 MHz and as high as 2000 MHz. Higher frequencies are used for achieving greater depths when locating groundwater tables, while lower frequencies reduce the impacts of surface roughness when determining soil moisture content. Two antennas measure the two-way travel time, which usually occurs in nanoseconds, because the wave velocity is high due to the low dielectric permittivity of most soils.

Soil types with higher water contents make locating the water table very difficult because the increased water content attenuates the signal much more quickly than for dry soil conditions. Attenuation values by material are provided in Table 6.1. In regions with a high clay content, the increase in water content throughout the capillary fringe affects the water table reflections because the signal attenuates much faster. The radar receives weaker signals with more dispersed characteristics so that the water table becomes less distinguishable (Doolittle et al. 2006). This can be circumvented if the radar is calibrated at measured depths. When the soil is primarily sand or gravel, the transition from vadose zone to water table happens more abruptly and provides a clearer image for the GPR. Table 6.1 shows the dielectric constants, conductivities, typical velocities, and attenuation rates from testing different materials using a specific GPR.

GPRs work best in coarse-grained soils, where the boundary between the unsaturated and saturated zones is very abrupt. Sands and gravels have extremely low magnetic properties and electrical conductivities. The soil particles do not typically retain much water, unlike clay particles, which have the cation-exchange capacity. The quick transition from the vadose zone to the water table means that the signal reflection back to the GPR is clearer, and the GPR can even estimate the depth with an accuracy of 20 cm (0.79 in.).

TABLE 6.1

Radar Characteristics for Various Materials Found in the Soils
(Adapted from Fisher et al. 1992)

Material	Dielectric Constant	Conductivity mSm^{-1}	Velocity mns^{-1}	Attenuation dB m^{-1}
Air	1	0	0.3	0
Distilled Water	80	0.01	0.033	0.002
Freshwater	80	0.5	0.033	0.1
Seawater	80	30000	0.01	1000
Dry Sand	3–5	0.01	0.15	0.01
Saturated Sand	20–30	0.1–1.0	0.1–1.0	0.03–0.3
Limestone	4–8	0.5–2	0.12	0.4–1.0
Shale	5–15	1–1000	0.09	1–100
Silts	5–30	1–1000	0.07	1–100
Clays	5–40	2–1000	0.06	1–300
Granite	4–6	0.01–1	0.13	0.01–1
Dry Salt	5–6	0.01–1	0.13	0.01–1
Ice	3–4	0.01	0.16	0.01

In clayey soils, the GPR signal attenuates much more rapidly due to the high water content and high electrical conductivity. The unique cation-exchange capacity and large surface area of clayey soils means that clay particles attract and hold more water than other soil types. The signal reflects off the water attached to the clay particle and has difficulty reaching lower layers where the groundwater table is located. Even though the signal may penetrate as deep as 30 m in coarse-grained soils such as sands and gravels, the signal typically goes down only a few meters in clayey soils. For soils containing more than 30% clay, the GPR is relatively ineffective. The GPR works best in soils with less than 10% clay content (Elkhetali 2006).

The four primary types of GPRs are single-offset, multi-offset, cross-borehole, and off-ground. Single- and multi-offset and borehole GPRs work in similar ways with transmitting and receiving antennas separated by a known distance (Figure 6.5). Off-ground GPRs are newer and combine the antennas into one unit that can be mounted on an all-terrain vehicle (ATV) or a vehicle and driven over the test area. Multi-offset GPRs have one transmitting antenna and multiple receiving antennas, which allow for data to be collected over a larger area. Two common acquisition geometries for multi-offset GPRs are common-midpoint (CMP) and wide angle reflection and refraction (WARR) (Figure 6.6). CMP acquisition keeps the transmitter at one common location while gradually increasing the distance between antennas. WARR acquisition gradually increases the distances between the antennas, including moving the transmitter. For many applications, results from multi-offset GPRs are required to use single-offset GPRs, but the process is time-consuming and expensive. Single- and multi-offset GPRs are very good for locating water table depth and soil moisture content.

When using cross-borehole GPRs, the antennas are lowered into two vertical boreholes. Under the zero-offset profile (ZOP) method, the two antennas are lowered in such a way that their midpoints are at the same depth. With multi-offset profile (MOP), the antennas are lowered in such a way that their midpoint depth varies in relation to the other so that the transmitter and receiver are not always even with each other. A schematic is provided in Figure 6.7. In the borehole, GPRs measure data vertically. This means that

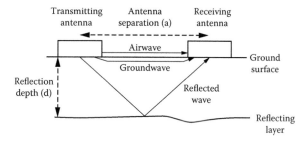

FIGURE 6.5
Schematic of single offset GPR to detect the depth to the water table (from Lunt et al. 2005).

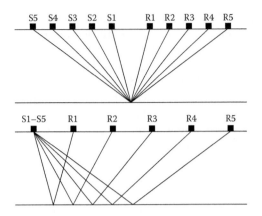

FIGURE 6.6
Schematic of common midpoint (top) and wide angle reflection and refraction (bottom) GPRs. S denotes the location of the transmitter, and R denotes the receiver (adapted from Huisman et al. 2003).

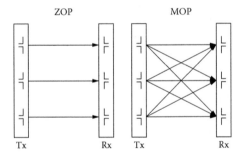

FIGURE 6.7
Schematic of zero-offset profile (ZOP) and multi-offset profile (MOP) borehole GPR showing direction from transmitter (Tx) to receiver (Rx) (adapted from Huisman et al. 2003).

borehole GPRs can collect data from separate layers, allowing for independent permittivities by layer instead of a relative permittivity that accounts for all the layers collected by offset GPRs. A disadvantage of borehole GPRs is that the GPRs can be inserted only at specific locations, whereas offset GPRs can more easily cover a large area.

The simplest equation to determine the depth to the water table, which uses the wave velocity and the two-way travel time is

$$d_w = \frac{vt_w}{2} \tag{6.4}$$

where d_w is the depth to the water table in m, v is the velocity of the radar signal in m sec^{-1}, and t_w is the two-way travel time in s. The two-way travel time is the elapsed time that it takes the wave to go from the transmitting antenna to the receiving antenna. Even though this equation is fairly simple, the wave velocity must often be calculated. The general velocity equation that applies to all soil types is controlled by the relative dielectric permittivity and the magnetic properties of the soil as follows:

$$v = \frac{c}{\sqrt{\varepsilon_r \mu_r \dfrac{1 + \sqrt{1 + \left(\dfrac{\sigma}{\omega\varepsilon}\right)^2}}{2}}} \quad (6.5)$$

where:
c = electromagnetic wave velocity in a vacuum = 3×10^{-8} m sec^{-1}
ε_r = relative dielectric permittivity of material
ε = dielectric permittivity in free space = 8.854×10^{-12} Fm^{-1}
μ_r = relative magnetic permeability
σ = electrical conductivity
ω = angular frequency of radar

There are no units for ε_r because it is a ratio of energy stored in a material to energy stored in a vacuum. The expression $\dfrac{\sigma}{\omega\varepsilon}$ is a loss factor, which for soils such as clean sand and gravel is approximately 0. In nonmagnetic materials (such as clean sand and gravel), μ_r equals 1. Therefore, the velocity equation in clean sands and gravel can be reduced to

$$v = \frac{c}{\sqrt{\varepsilon_r}} \quad (6.6)$$

With a single-offset GPR, the velocity can be approximated if the depth to the water table or reflecting layer is known:

$$v = \frac{2\sqrt{x^2 + d^2}}{t_w} \quad (6.7)$$

where:
x = position relative to the reflecting layer
d = depth to reflecting layer
t_w = two-way travel time

If the two antennas are separated by a significant distance, this should be incorporated into the velocity equation:

$$v = \frac{\sqrt{(x - 0.5a)^2 + d^2} + \sqrt{(x + 0.5a)^2 + d^2}}{t_w} \quad (6.8)$$

where a is the distance between antennas. If the velocity is known, Equation (6.6) can be inverted to solve for the effective dialectric permittivity:

$$\varepsilon = \left(\frac{c}{v}\right)^2 \tag{6.9}$$

To determine the volumetric water content, the following equation can be inverted:

$$\varepsilon = \left[(1-\eta)\sqrt{\varepsilon_s} + (\eta - VWC)\sqrt{\varepsilon_a} + VWC\sqrt{\varepsilon_w}\right]^2 \tag{6.10}$$

where:
ε = effective dielectric permittivity
η = soil porosity
VWC = free soil water content
ε_s = dielectric permittivity of soil
ε_a = dielectric permittivity of air ≈ 1
ε_w = dielectric permittivity of water ≈ 80

When inverted, the volumetric water content is equal to

$$VWC = \frac{(1-\eta)\sqrt{\varepsilon_s} + \eta\sqrt{\varepsilon_a} - \sqrt{\varepsilon}}{\sqrt{\varepsilon_a} - \sqrt{\varepsilon_w}} \tag{6.11}$$

If the gravimetric water content (GWC) is desired, a simple conversion using the bulk density can be used as

$$GWC = \frac{VWC}{\rho_d} \tag{6.12}$$

where ρ_d is the bulk density of the soil.

6.4 Subsurface Water

The hydrologic cycle describes the location and volume of water in a watershed or distributed across the globe. Measuring the various components of the hydrologic cycle are critical for managing Earth's freshwater resources and for monitoring climate change. In situ measurements are often difficult to obtain in remote regions and always suffer from their point-to-area representativeness. Radar remote sensing provides for the measurement of hydrologic cycle states and fluxes at unprecedented spatiotemporal resolutions across the globe. A pulsed Doppler radar can measure the surface velocity, while the same radar operating at UHF can measure the river stage and the channel's cross-section. This information can be combined to estimate river discharge using noncontact methods. Satellites carrying radars, such as the KaRIN proposed for the SWOT mission, can be used to measure surface water heights for inland water bodies and oceanic regions to a precision of 1 cm. SARs onboard satellites provide surface water extent and inundation stemming from river floods and storm surges. Soil moisture in the top 0–5 cm and even root-zone soil moisture can be measured by C-, L-, and P-band radars from space. Finally, subsurface water can be detected with ground-penetrating radars. They can retrieve soil moisture at deeper layers, although they perform best in low-clay-content soils. GPRs can also be used to determine the depth to the groundwater table.

These new remote-sensing technologies will offer unique observations in the coming decades that will reshape theories and mathematical formulas describing the complex movement and storage of water within Earth's system. Ultimately, these concepts and observations will be encompassed in models that will predict the hydrologic cycle components, leading to better practices for sustainability.

Problem Sets

QUALITATIVE QUESTIONS

1. How can radars help estimate the variables in the water balance equation? How do they affect the measurements in the balance equation?

2. How can radars help estimate the variables in the water budget equation? How do they affect the measure budget?

3. What information will the future SMAP mission provide that is not currently obtained with the previous and current missions?

4. Why would it be useful to know the soil moisture for only the top few centimeters of soil, using ASCAT and SMAP?

5. When would offset GPRs be preferable to use in place of borehole GPRs? When would the opposite be true?

6. Since most soils throughout the United States are clayey soils, what use are GPRs in the United States?

7. How do different materials affect the results from GPRs? Also compare similar materials. Discuss the reasoning for deciding which materials are similar, i.e., which properties determine similarity? Use Table 6.1 as a reference.

QUANTITATIVE QUESTIONS

1. (a) If the relative dielectric constant, ε, of a soil is 6, what is the electromagnetic wave velocity? (b) If the travel time is 39.8 ns, how deep is the water table?

2. (a) If the dielectric constant of a soil, ε_s, is 8, the effective porosity is 0.15, and the volumetric water content is 0.5, what is the effective dielectric permittivity? (b) If the specific gravity of the soil is 2.7, what is the gravimetric water content?

References

Bolten, J. D., V. Lakshmi, and E. Njoku. 2003. Soil moisture retrieval using the passive/active. *IEEE Transactions on Geoscience and Remote Sensing* 41 (12): 2792–2812.

Brocca, L., S. Hasenauer, T. Lacava, F. Melone, T. Moramarco, W. Wagner, et al. 2011 August. Soil moisture estimation through ASCAT and AMSR-E sensors: An intercomparison and validation study across Europe. *Remote Sensing of Environment* 115: 3390–3408.

Colliander, A., S. Chan, S.-B. Kim, N. Das, S. Yueh, M. Cosh, et al. 2012. Long term analysis of PALS soil moisture campaign measurements for global soil. *Remote Sensing of Environment* 121: 309–322.

Costa, J. E., R. T. Cheng, F. P. Haeni, N. Melcher, K. R. Spicer, E. Hayes, W. Plant, K. Hayes, C. Teague, and D. Barrick. 2006. Use of radars to monitor stream discharge by noncontact methods. *Water Resources Research* 42, W07422.

Creutin, J. D., M. Muste, A. A. Bradley, S. C. Kim, and A. Kruger. 2003. River gauging using PIV techniques: A proof of concept experiment on the Iowa River. *Journal of Hydrology* 277 (3–4): 182–194.

Di Baldassarre, G., G. Schumann, and P. D. Bates. A technique for the calibration of hydraulic models using uncertain satellite observations of flood extent, *Journal of Hydrology* 367, 276.

Doolittle, J. A., B. Jenksinson, D. Hopkins, M. Ulmer, and W. Tuttle. 2006. Hydropedological investigations with ground-penetrating radar (GPR): Estimating water-table depths and local ground-water flow pattern in areas of coarse-textured soils. *Geoderma* 317–329.

Elkhetali, S. 2006. Detection of groundwater by ground penetrating radar. *Progress in Electromagentics Research Symposium (PIERS)*, pp. 251–255. Cambridge.

Entekhabi, D., et al. 2010. The soil moisture active passive (SMAP) mission. *Proceedings of the IEEE* 98 (5): 704–716.

Fisher, E., G. A. McMechan, and A. P. Annan. 1992. Acquisition and processing of wide-aperature ground-penetrating radar data. *Geophysics* 57 (3), 495–504.

Hong, Y., S. I. Khan, C. Liu, and Y. Zhang. 2012. Global soil moisture estimation using microwave remote sensing. In *Multiscale Hydrologic Remote Sensing: Perspectives and Application*, Ni-Bin Chang and Yang Hong, Eds. CRC Press, 399–410.

Huisman, J. A., S. S. Hubbard, J. D. Redman, and A. P. Annan. 2003. Measuring soil water content with ground penetrating radar: A review. *Vadose Zone* 2: 476–491.

Lunt, J. A., S. S. Hubbard, and Y. Rubin. 2005. Soil moisture content estimation using ground-penetrating radar reflection data. *Journal of Hydrology* 307: 254–369.

Mason, D. C., P. D. Bates, and J. T. Dall'Amico. 2009. Calibration of uncertain flood inundation models using remotely sensed water levels. *Journal of Hydrology* 368, 224–236.

Matgen, P., G. Schumann, J.-B. Henry, L. Hoffmann, and L. Pfister. 2007. Integration of SAR-derived inundation areas, high precision topographic data and a river flow model toward real-time flood management. *International Journal of Applied Earth Observation and Geoinformation* 9, 247–263.

Schumann, G. J.-P., J. C. Neal, D. C. Mason, and P. D. Bates. 2011. The accuracy of sequential aerial photography and SAR data for observing urban flood dynamics, a case study of the UK summer 2007 floods. *Remote Sensing of Environment* 115 (10), 2536–2546.

Simpson, M. R. and R. N. Oltmann. 1993. Discharge measurement using an acoustic Doppler current profiler: U.S. Geological Survey Water-Supply Paper 2395, 34 pp.

Stephens, E. M., P. D. Bates, J. E. Freer, and D. C. Mason. 2011. The impact of uncertainty in satellite data on the assessment of flood inundation models. *Journal of Hydrology*, 414–415, 162–173.

Tralli, D. M., R. G. Blom, V. Zlotnicki, A. Donnellan, and D. L. Evans. 2005. Satellite remote sensing of earthquake, volcano, flood, landslide and coastal inundation hazards. *ISPRS Journal of Photogrammetry and Remote Sensing* 59: 185–198.

Yorke, T. H., and K. A. Oberg. 2002. Measuring river velocity and discharge with acoustic Doppler profilers. *Flow Measurement and Instrumentation* 13: 191–195.

Yuen, K. 2012. *SMAP: Soil Moisture Active Passive.* Retrieved October 2012 from NASA Jet Propulsion Laboratory: http://smap.jpl.nasa.gov/.

7

Radar QPE for Hydrologic Modeling

One of the most ubiquitous uses for radar data has been in hydrologic modeling. Radars provide precipitation estimates at a space-time resolution that is sufficient to forecast flash floods when input to a hydrologic model. Moreover, radars estimate properties of the land surface that can be used in model parameterization and state variable estimation. Radars can even measure streamflow, which is the primary output from a hydrologic model and thus can be used to evaluate forecasts from a hydrologic model. This chapter details the basics of hydrologic modeling with a focus on those applications and models that best utilize radar observations. Techniques that use models to perform hydrologic evaluations of precipitation inputs are also introduced. This provides for an evaluation framework that identifies the best approaches to precipitation estimation based on the hydrologic perspective of the model outputs, rather than the conventional radar rainfall-to-gauge comparisons. In practice, both methods are employed in tandem.

7.1 Overview of Hydrological Models

7.1.1 Model Classes

Hydrologic models come in many different configurations for simulating the water balance. This section provides a conceptual overview of general hydrologic model characteristics and provides some specific examples. Figure 7.1 shows the basic structure of a hydrologic model. The first defining characteristic of a hydrologic model is the degree of physical realism contained in the equations in the **model structure**, f. Simplified models consist of parameterized processes and require observations of system behavior ($O(x,t)$) to estimate the model parameters ($W(x)$), which may have no physical meaning. In the case of soil storages and fluxes, the concepts contained in a simple model often resemble that of a bucket. The bucket has a given storage capacity. When this capacity is exceeded from input rainfall from above, the bucket "spills" and streamflow is generated on the surface. Obviously, modeling the hydrologic cycle demands much more sophisticated approaches than this, but simpler **conceptual models** are useful for applications where great computational expense cannot be afforded. Examples include testing

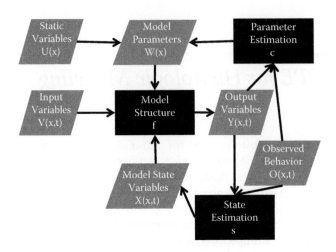

FIGURE 7.1
Basic components of a hydrologic modeling system. The gray boxes denote the static and dynamic variables, while the black boxes contain functional relationships (equations).

the performance of complex data assimilation and parameter estimation methods that might require thousands or even millions of iterative simulations. It may be infeasible to run a very complex hydrologic model at high resolution when minimizing an objective function, as is typically done in state and parameter estimation. Simpler conceptual models are more suitable for running over large spatial domains and for long time periods spanning years or even decades. Moreover, they require less knowledge of the composition of the soils, vegetation, and other characteristics of the basin. Some of these data sources may not exist, which requires parameterization rather than an explicit representation of physical processes.

More complex hydrologic models, often referred to as **physically based models**, use more variables and thus require more functional relationships (equations) to interrelate the variables. It should be noted that no real defining distinction exists between a conceptual and a physically based model. All models have some degree of conceptualization built into them. Even if one were to employ a model with explicit 3-D equations based on physical principles with no parameterization at all, then there must be a numerical approximation when solving the model equations. Thus, the transition from conceptual models to so-called physically based models is really a continuum of physical realism contained within the model equations and parameters. Models on both ends of the spectrum have advantages and disadvantages in terms of performance, data requirements, spatiotemporal resolution, and resulting outputs. Despite a greater computational expense, complex, physically based models offer the capability of providing outputs in locations with no historical measurements of observed system behavior (i.e., streamflow). This is an issue addressed more fully in Chapter 8.

The next defining characteristic of a hydrologic model is its discretization of inputs, equations, parameters, and outputs. Simplified, conceptual models require observations of input and outputs (e.g., rainfall, temperature, and streamflow) to estimate model parameters, which limits their applicability to the point (or basin outlet) where the streamflow observations are available. One end of the spectrum of spatial discretization is basin-wide treatment of all processes; this is a **lumped model**. A lumped model uses a single value for rainfall input that is the basin-averaged rainfall. These models were some of the first ones to be designed and implemented in operational forecast systems. If the historical datasets describing the system behavior are accurate and complete, then these lumped models can forecast streamflow with great accuracy. The estimated parameter values do not readily transfer to smaller, subbasins nested in the parent basin, nor do they directly transfer to adjacent basins. Moreover, these models find difficulties in forecasting streamflow for events that fall out of their training dataset, such as with extreme flooding events. They assume forecast outputs will resemble those that occurred in the past given the same rainfall inputs and soil moisture conditions. Generally, this is a good assumption, but it can fail for basins whose behavior has been altered due to advertent and inadvertent anthropogenic impacts such as urbanization and climate change, respectively.

A fully **distributed**, or raster-based, model solves for water balance at each grid point. These models often have their conceptual basis of runoff production from a conceptual, lumped model. Others employ the Saint Venant equations for modeling overland and channel flow while using Richards' equation and Darcy's law for unsaturated and saturated flow, respectively. These equations are solved in a finite-difference or finite-element manner across a grid of elements describing the basin. The grid cells are hydrologically connected using information from a digital elevation model (DEM). Because observations of streamflow are rarely available at each of the grid points in a distributed model's domain, they must rely on relationships between model parameters and observable features of the land surface such as soil types, soil depths, and land cover. Koren et al. (2000) provide an example of estimating parameters for a distributed conceptual model using physical soil properties. A basin may also be subdivided into several smaller catchments, which are independently modeled and used as input to a downstream catchment. This approach is referred to as **semidistributed** modeling.

The structure of the Coupled Routing and Excess STorage model (CREST) described in detail in Wang et al. (2011) is illustrated in Figure 7.2. Each compartment represents a storage component, whether it be above the surface on the vegetation canopy or one of the three soil layers, and the arrows connecting each storage tank represent fluxes. The diamonds correspond to partitions that are generally dictated by thresholds. If we are to follow a raindrop, P, we see that it becomes intercepted by the vegetation canopy first, and some of the water is evapotranspired via E_c back to the atmosphere. Precipitation that makes it to the soil layer P_{soil} may become surface runoff

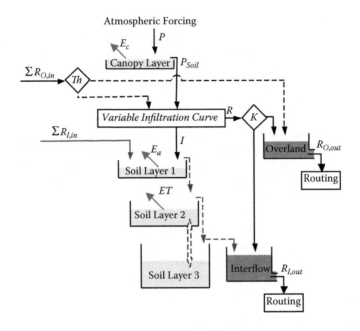

FIGURE 7.2
The conceptual structure of the distributed Coupled Routing and Excess STorage model (CREST) described in detail in Wang et al. (2011).

R immediately, depending on the characteristics of the variable infiltration curve (VIC; also called **tension water capacity curve**) originally founded in the Xinanjiang model (Zhao et al. 1980, 1992) and later employed in the University of Washington VIC model (Liang et al. 1996). The curve describing variable infiltration follows as

$$i = i_m \left[1 - \left(1 - \frac{W}{W_m} \right)^{\frac{1}{1+b}} \right] \tag{7.1}$$

where i is the point infiltration capacity; i_m is the maximum infiltration capacity of a cell; and b is the exponent of the curve. i_m is a function of the cell's maximum water capacity (W_m) of the three soil layers as

$$i_m = W_m (1 + b) \tag{7.2}$$

$$W_m = W_{m1} + W_{m2} + W_{m3} \tag{7.3}$$

and

$$W = W_1 + W_2 + W_3 \tag{7.4}$$

where W_1, W_2, and W_3 are the cell's mean soil water depth of each layer. The amount of water available for infiltration (I) is then computed as follows:

$$\text{When} \quad i + P_{Soil} \geq i_m \quad I = (W_m - W)$$

$$\text{When} \quad i + P_{Soil} < i_m \quad I = (W_m - W) + W_m \left[1 - \frac{i + P_{Soil}}{i_m}\right]^{1+b} \quad (7.5)$$

The soil storage reservoirs are filled in sequence from top to bottom. Surface runoff occurs when no more water is available for infiltration I. This can be caused by a combination of heavy precipitation rates P_{soil} that overwhelm the infiltration capabilities of the soils i_m and soils that are saturated from prior rainfall and are unable to store any more water; the generation of these surface overland flows are often referred to as infiltration excess (Hortonian) and saturation excess (Dunne), respectively. It should be noted that the CREST model enables water that has already been designated as overland flow R_O from upstream grid cells to reenter the soils and potentially become infiltrated soil water. This is an approximation for the natural process that occurs in losing rather than gaining streams.

As soon as surface runoff R is generated, it is further partitioned into overland R_O or interflow R_I depending on the soil's saturated hydraulic conductivity values (K_s). The difference between these channel flows can be thought of as quick-flow responses to rainfall events and then slow-flow, or baseflow. Note that R_I enters the soil tanks in grid cells downstream and can thus act to fill the soil layers leading to additional soil saturation and subsequent surface runoff. The connectivity, speed, and direction of flows are dictated by a digital elevation model (DEM) that has been processed using geographic information systems. These steps ensure the DEM is suitable for use in a hydrologic modeling application. This means the DEM and its derivatives must be postprocessed to ensure that no relative minima are in the elevation values along the downstream direction. These dips are called sinks and will cause water to pond instead of continue its flow downstream. Therefore, the postprocessing step of "filling the sinks" adds values to the sinks so that water will continue to flow downstream. Furthermore, channels can be forced into the DEM by artificially removing elevation values at the known channel locations. This process is called **burning in the streams**.

The storage capacities of each of the tanks in the CREST model shown in Figure 7.2 as well as the fluxes connecting them must be parameterized. CREST is a distributed parameter model, so the parameter values vary from cell to cell. Thus, instead of relying on observations of rainfall and runoff to estimate model parameters, as is detailed below, more reliance is placed on relating the parameters to measurable, physical properties of the land surface. The advent of remote-sensing technologies from space has provided a wealth of information about Earth's hydrologic properties.

Typically, datasets such as soil types and land cover are gridded and globally available, and thus readily accommodate distributed hydrologic models. However, the relationship between hydrologic model parameters and the variables measured from space can be indirect or approximate. Moreover, parameters may be in the model that have little resemblance to physical processes or may be very difficult to measure. Nonetheless, the CREST distributed model, and many others like it, come with an a priori database of gridded parameters. This enables the model to be run uncalibrated, i.e., without a lengthy parameter estimation period, and produce reasonable results. But, as we'll see in the next section, observations describing the system behavior can be used to improve the model simulations through parameter estimation.

7.1.2 Model Parameters

Regardless of whether a model is conceptual or physical, it will have a variety of **model parameter values** ($W(x)$) for controlling and adjusting how water propagates inside the model. Examples of parameters can include values that control the amount of water that will infiltrate under saturated conditions, surface roughness that controls the speed at which water moves through the channel network, and the relationship between channel area and channel streamflow. To provide the best possible model output, the parameters are often calibrated in an optimization procedure, which seeks to minimize the error between the model simulations and observed values by exploring the available parameter space. This process c is illustrated in Figure 7.1 by comparing the output variables ($Y(x,t)$) with the observed system behavior ($O(x,t)$). If they disagree, then the parameter values $W(x)$ are adjusted and the model is rerun with the same input variables $V(x,t)$. The process continues until the simulations match the observations according to some objective criteria. This iterative procedure is complicated by model parameters that interact, by local (not global) optima in the multidimensional parameter space, and by parameters that have sensitivity only under specific conditions.

Model parameter estimation is generally designed to be performed offline for a given model, observational datasets, and basin. Following this calibration procedure, the model parameters are fixed and the model can be used in forecast mode. It is recommended to evaluate the model simulations using an independent dataset during a validation phase. Parameter estimation can be done manually or automatically. Manual calibration is a useful learning exercise in order to experience the various controls the model parameters have on hydrologic simulation and to illuminate how the parameters interact. Automatic methods are often more efficient and useful for operational applications. Shuffled complex evolution (Duan et al. 1993) and Differential Evolution Adaptive Metropolis (DREAM; Vrugt et al. 2009) are two common automatic optimization methods used for parameter estimation. These methods have been designed to explore the multidimensional parameter space through multichain Monte Carlo sampling. There is often considerable uncertainty in the exact values of these parameters as the optimization process can compensate for errors in the observational datasets or even in the model equations. For this reason, optimized parameters are often conditioned on the quality of the observational datasets that were incorporated during the calibration procedure. If the observations or model formulations change, then the parameter values need to be updated. Clearly, this procedure doesn't accommodate new remote-sensing technologies and algorithms that are continually evolving.

7.1.3 Model State Variables and Data Assimilation

Model **state variables** ($X(x,t)$) typically consist of soil moisture and amount of water retained in storages near the surface and underground as well as water in the channels. State variables are important for their use in evaluating the performance of a hydrologic model. Streamflow is often directly described or derived from a set of state variables. Estimating the state variables in a hydrologic model poses some challenges. One approach is to "warm up" the soil and stream state variables by forcing the hydrologic model with observed precipitation and temperature for a reasonably long time period (typically, months) leading up to the time at which a forecast is desired. This method works on the principle that a long enough time period will allow the model states to come into equilibrium that represents the true nature of the states. This is important to keep in mind when working with radar precipitation estimates because as new algorithms are developed, there must also be reanalyses of precipitation in order to provide a continuous, long, and consistent precipitation record necessary to warm up a hydrologic model.

As we saw in Chapter 6, radar remote-sensing methods have provided new observations of soil states and fluxes near the surface and below. Another tactic for estimating state variables, represented by the process *s* in Figure 7.1, is to incorporate these observations that are related to soil and river states into a hydrologic model **data assimilation** framework. The basic idea is to adjust model state variables based on the observations in order to improve the physical realism of the model and to improve the accuracy of forecast variables, namely streamflow. Below, we highlight techniques to update model states and so the term *data assimilation* hereafter refers specifically to this topic. The same term can also apply to smoothing and filtering methods. Common algorithms for state updating, namely *Kalman filters, variational methods* (i.e., *3DVAR* and *4DVAR*), and *ensemble-based techniques,* are covered herein.

The **Kalman filter** (KF; Kalman 1960) is a sequential filtering method based on the minimum variance or least squares framework. The basic assumptions in KF are the normality of error distributions and linearity of error growth (Hamill 2006), and that the expected values of the errors from both the model and the observations are unbiased and not correlated.

It contains two steps: the forecast step and the assimilation step. In the forecast step, the model is run using previous information to generate the forecast and its error covariance:

$$x_i^b = M_{i-1} x_{i-1}^a \tag{7.6}$$

$$P_i^b = M_{i-1} P_{i-1}^a \left(M_{i-1}\right)^T + Q_{i-1} \tag{7.7}$$

where x_i^b is the state background forecast or first guess, M_{i-1} represents the linear model that advances from time step i–1 to i given the analysis data x_{i-1}^a, P_i^b is the background error covariance, P_{i-1}^a is the analysis error covariance from the previous time, and Q_{i-1} is the model error. In the assimilation step, the optimal estimate (or analysis) and its error covariance are obtained by updating the state background forecast using the observation and its error information:

$$x_i^a = x_i^b + K_i \left(y_i - H_i x_i^b\right) \tag{7.8}$$

$$K_i = P_i^b H_i^T \left(H_i P_i^b H_i^T + R_i\right)^{-1} \tag{7.9}$$

$$P_i^a = \left(I - K_i H_i\right) P_i^b \tag{7.10}$$

where x_i^a is the analysis with error covariance P_i^a at time i, K_i is the weight applied to the innovation term $\left(y_i - H_i x_i^b\right)$ called the **Kalman gain**, H_i is the linear operator that converts states into observation space, y_i is the observation with error covariance R_i, and I is the identity matrix.

The **extended Kalman filter** (EKF; Jazwinski 1970) is implemented for nonlinear models. The same equations described above for the KF applies, with the following modifications: (1) The term M_{i-1} in Equation (7.6) is the nonlinear model, while the same term in Equation (7.7) refers to the linearized version of the model, and (2) the observation operator H_i is linear for Equation (7.8) and nonlinear for Equations (7.9) and (7.10). A further development on the Kalman filter that addresses the nonlinearity challenges is the **ensemble Kalman filter** (EnKF; Evensen 1994), which is described in subsequent sections in this chapter.

Variational methods are techniques based on the maximum likelihood or Bayesian framework where estimates of the states are used to minimize a cost function of the form:

$$J_i(x_i) = \frac{1}{2}\left\{\left[y_i - H_i(x_i)\right]^T R_i^{-1}\left[y_i - H_i(x_i)\right] + \left(x_i^b - x_i\right)^T \left(P_i^b\right)^{-1}\left(x_i^b - x_i\right)\right\} \tag{7.11}$$

where $x_i = x_i^a$ is the value that minimizes J_i. x_i^a is found through numerical minimization in an iterative procedure:

Step 1: Start with the first guess $x_i = x_i^b$ and compute the cost function J_i.

Step 2: Compute the gradient of J_i with respect to x_i $(\nabla J_{x_i,i})$.

Step 3: Using an optimization algorithm such as the conjugate gradient method, and the values of J_i and $J_{x_i,i}$ computed in Steps 1 and 2 respectively, determine the correction for x_i.

Step 4: Check for convergence by computing the norm of the gradient $\nabla J_{x_i^{new},i}$: if convergence has been reached, the value of the norm of the gradient of J_i with respect to the corrected value of x_i (i.e., x_i^{new}) must be approximately zero. If convergence has not been reached, repeat the procedure starting with x_i^{new}.

The above procedure is what has come to be known as **three-dimensional variational method** (3DVAR). Figure 7.3 presents an example of the application of the iterative procedure in 3DVAR for a two-variable problem.

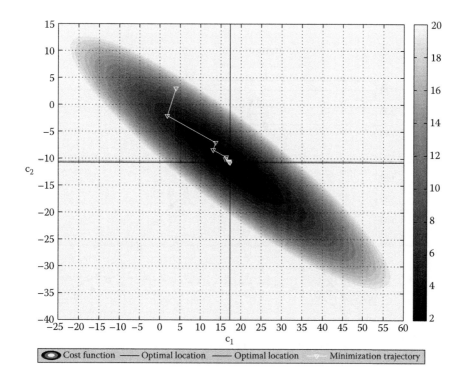

FIGURE 7.3
Illustration of the iterative procedure used in 3DVAR for minimizing the cost function in Equation (7.11).

The **four-dimensional variational method (4DVAR)** is applied when multiple observations distributed in time are available and the assimilation of data is performed for the interval defined by these observations. In this case, the objective is to find the value of the states at the beginning of the interval (i.e., $i = 0$) that minimizes the cost function defined for the entire assimilation window:

$$J(x_0) = \frac{1}{2} \sum_{i=0}^{t} \left\{ \left[y_i - H_i(x_i) \right]^T R_i^{-1} \left[y_i - H_i(x_i) \right] + \left(x_0^b - x_0 \right)^T \left(P_0^b \right)^{-1} \left(x_0^b - x_0 \right) \right\} \quad (7.12)$$

where t is the number of time steps at which observations are available. The procedure to minimize the cost function is similar to that of 3DVAR. However, 4DVAR requires the computation of the linear tangent or adjoint model (i.e., a first-order approximation of the model trajectory). Figure 7.4 depicts the implementation of 4DVAR as compared with 3DVAR.

The EnKF is a Monte Carlo simplification of the EKF. The most important advantage of EnKF is that error statistical information is retrieved from the ensembles, and thus the linearized model and observation operator in Equations (7.7) and (7.8) are not necessary. Figure 7.5 presents a schematic

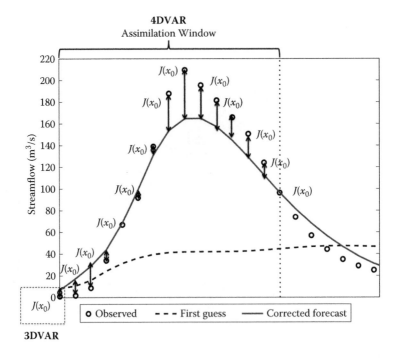

FIGURE 7.4
Illustration of 4DVAR cost function minimization in Equation (7.12) as a function of time.

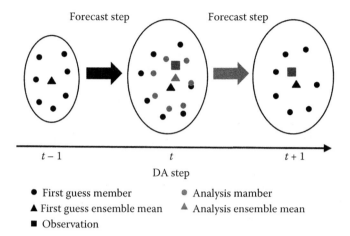

Forecast step   Forecast step

$t-1$ t $t+1$

DA step

- First guess member   • Analysis mamber
▲ First guess ensemble mean   ▲ Analysis ensemble mean
■ Observation

FIGURE 7.5
Illustration of data assimilation using the ensemble Kalman filter.

of the ensemble data assimilation procedure. There are two approaches in EnKF: stochastic and deterministic (Hamill 2006). The difference between the two is that observations are perturbed in the stochastic approach by adding noise from a normal distribution with zero mean and standard deviation R, while in the deterministic method the observations are not perturbed. Perturbations of the observations are necessary because otherwise the error covariance of the analysis is systematically underestimated (Hamill 2006). However, the noise added to the observations can have a detrimental effect (Clark et al. 2008). Whitaker and Hamill (2002) developed a deterministic EnKF version entitled the **ensemble square root filter** (EnSRF). The EnSRF uses a reduced Kalman gain to update the perturbations. The equations for this implementation are very similar to those described for the general KF application. The forecast step is given by

$$x_{i,k}^{b} = M_{i-1,k}\left(x_{i-1,k}^{a}\right) \qquad for \ k = 1,2,3,...,L \tag{7.13}$$

$$\bar{x}_{i}^{b} = \frac{1}{L}\sum_{k=1}^{L} x_{i,k}^{b} \tag{7.14}$$

$$x_{i,k}^{\prime b} = x_{i,k}^{b} - \bar{x}_{i}^{b} \tag{7.15}$$

where L is the ensemble size, $x_{i,k}^{b}$ is the kth member of the background ensemble, $\bar{x}_{i}^{b}$ is the mean of the background ensemble, and $x_{i,k}^{\prime b}$ is the perturbation at time i. As for the background error covariance, its value is calculated from the ensemble. However, instead of computing and storing P_i^b, $P_i^b H_i^T$, and $H_i P_i^b H_i^T$, the following is calculated (Whitaker and Hamill 2002):

$$P_i^b H_i^T \approx \frac{1}{L-1} \sum_{k=1}^{L} \left(x_{i,k}^b - \bar{x}_i^b \right) \left[H_i(x_{i,k}^b) - \bar{H}_i(x_{i,k}^b) \right]^T \tag{7.16}$$

$$H_i P_i^b H_i^T \approx \frac{1}{L-1} \sum_{k=1}^{L} \left[H_i(x_{i,k}^b) - \bar{H}_i(x_{i,k}^b) \right] \left[H_i(x_{i,k}^b) - \bar{H}_i(x_{i,k}^b) \right]^T \tag{7.17}$$

The data assimilation step is divided into mean update and the perturbation update. The mean update:

$$\bar{x}_i^a = \bar{x}_i^b + K_i \left[y_i - \bar{H}_i(x_{i,k}^b) \right] \tag{7.18}$$

$$K_i = P_i^b H_i^T \left[H_i P_i^b H_i^T - R_i \right]^{-1} \tag{7.19}$$

where K_i is the traditional Kalman gain. Now, the perturbation update:

$$x_{i,k}'^a = x_{i,k}'^b - \tilde{K}_i H_i'(x_{i,k}^b) \tag{7.20}$$

$$\tilde{K}_i = \left(1 + \sqrt{\frac{R_i}{H_i P_i^b H_i^T + R_i}} \right)^{-1} K_i \tag{7.21}$$

where $\tilde{K}_i$ is the reduced Kalman gain. It can be seen that the perturbations are reduced less with $\tilde{K}_i$ than with K_i, yielding the same effect as with the EnKF with perturbed observations. The final analyses are then computed by

$$x_{i,k}^a = \bar{x}_i^a + x_{i,k}'^a \tag{7.22}$$

It should be noted that the fundamentals of the process of assimilating observations are independent from their source. Streamflow measurements represent the flow of water in a channel, whether they come from stream gauges, radar, or satellite signals. Furthermore, any given data assimilation algorithm can operate independently from the source of the observations. However, in practice it is important to have information about the errors associated to the derivation of the observations, which directly relates to their source. For example, from Chapter 6, we saw measurements from a ground penetrating radar are highly sensitive to the conductivity of water. In this case, the error covariance term will be a function of the water conductivity.

Streamflow is arguably the most important variable in the hydrologic modeling field, perhaps because it accounts for the integrated processes of water fluxes through a watershed. It is in most cases the primary output of hydrologic models, which makes the application of assimilation techniques relatively straightforward. Streamflow is usually measured at specific locations in a watershed such as its outlet. In some regions, observations are available at multiple points in a single basin from gauge measurement networks. The manner in which streamflow observations are distributed in space has implications on the implementation of the assimilation technique. Since streamflow results from the space-time integration of hydrologic processes occurring at multiple points within a basin (e.g., rainfall or evapotranspiration), assimilation of streamflow measurements needs to account for the spatiotemporal structure of the covariance between model states and outputs at any given location. This particular consideration can be neglected for the case of lumped hydrologic models, where all variables in the modeling system are regarded as spatially invariant and treated as a single point.

The EnKF was employed to assimilate hourly streamflow observations into a simple but widely used conceptual rainfall-runoff model for flood prediction purposes. The hydrologic model is a lumped, conceptual model represented by a nonlinear tank, based on the probability distributed model developed by Moore (1985), whose output is routed through two series of linear tanks denoting quick and slow flow components of the streamflow. Errors in streamflow observations were characterized through a simple empirical model. The model assumes that error in streamflow observations grows log-linearly as a function of streamflow magnitude as shown in Figure 7.6. The model was implemented on the Tar-Pamlico River basin in coastal North Carolina for this work (Figure 7.7). The catchment has a drainage area of 5709 km² and is located on the coastal plain. The EnSRF was used to perform data assimilation of streamflow observations for an event in

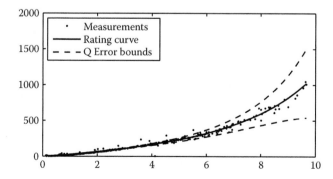

FIGURE 7.6
Error model for the rating curve used to estimate streamflow from stage height measurements.

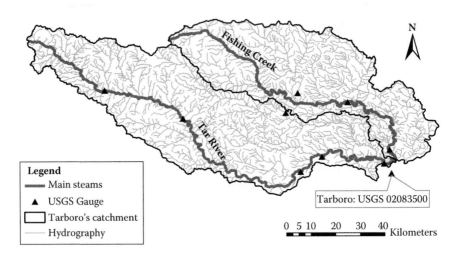

FIGURE 7.7
The Tar-Pamlico River basin in coastal North Carolina. The drainage area for the entire basin is 5709 km².

March of 2003 (Figure 7.8). The hydrographs and error metrics clearly show that the simulation of streamflow dramatically improves when data assimilation is utilized. Furthermore, the simulation falls within the observation uncertainty bounds.

7.1.4 Hydrological Model Evaluation

The accuracy of streamflow forecasts can be improved through better observations of precipitation and temperature, improved model parameters, better estimates of state variables, and more accurate and descriptive model physics. To make improvements to a model, we must be able to quantify its accuracy conditioned on the improvements being tested; this is the basis of **hypothesis testing**. Hydrologic model evaluation is also what is inherently taking place in model parameter estimation methods previously discussed. Evaluation of a hydrologic model comes in the form of comparing simulated variables such as streamflow and soil moisture to observed values. Evaluations with soil moisture can be performed at a single location (i.e., where there is an in situ sensor) by comparing the time series of observations to simulations. Similarly, it is now becoming possible to compare spatially distributed variables to observations. Flood inundation extent is another observation that can be used to evaluate and improve a hydrologic model as has been demonstrated for urban hydraulic modeling applications (Schumann et al. 2011). A distributed hydrologic model is more applicable in these contexts because satellite-based soil moisture and inundation products are spatially distributed (gridded), thus making it feasible to compare state variables to observations at individual points.

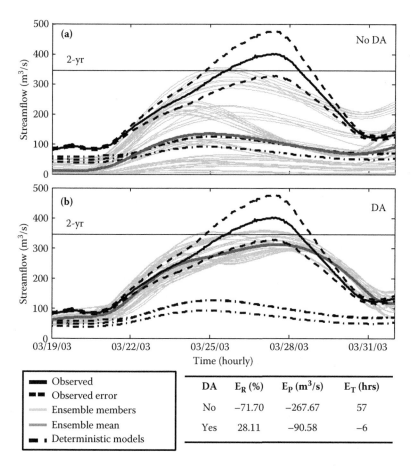

DA	E_R (%)	E_P (m^3/s)	E_T (hrs)
No	−71.70	−267.67	57
Yes	28.11	−90.58	−6

Legend:
- Observed
- Observed error
- Ensemble members
- Ensemble mean
- Deterministic models

FIGURE 7.8
Time series of simulated streamflow on the Tar-Pamlico River using HyMOD with (a) no data assimilation and (b) assimilation of streamflow observations using EnSRF. The statistics in the embedded table are for the relative error (in percentage) of time-integrated runoff (E_R), absolute error (in m^3 s^{-1}) in peakflow simulation (E_P), and absolute error (in hrs) in simulating the time at which peakflow occurred (E_T). All errors are substantially reduced following implementation of the EnSRF data assimilation procedure.

Streamflow is the most common variable that is evaluated with statistics such as the Nash–Sutcliffe efficiency (NSE; Nash and Sutcliffe 1970):

$$\text{NSE} = 1 - \frac{\sum_{i=1}^{n}(Q_i^{obs} - Q_i^{sim})^2}{\sum_{i=1}^{n}(Q_i^{obs} - Q_{mean}^{obs})^2} \tag{7.23}$$

where Q_i^{obs} is the ith observed value, Q_i^{sim} is the ith simulated value, and Q_{mean}^{obs} is the arithmetic mean of the observed values. This score compares the skill of the model to the skill of using the average observed streamflow. Subsequently good skill is defined as performing better than the average of the observed streamflow, resulting in an NSE > 0. NSE values < 0 implicate the model performs worse than the Q_i^{obs}, and a value of 1 is for a simulation that matches the observed perfectly at each time step.

The linear sample Pearson correlation coefficient:

$$r = \frac{\sum_{i=1}^{n}(Q_i^{obs} - Q_{mean}^{obs})(Q_i^{sim} - Q_{mean}^{sim})}{\sqrt{\sum_{i=1}^{n}(Q_i^{obs} - Q_{mean}^{obs})^2}\sqrt{\sum_{i=1}^{n}(Q_i^{sim} - Q_{mean}^{sim})^2}} \tag{7.24}$$

where Q_i^{obs} is the ith observed value, Q_i^{sim} is the ith simulated value, Q_{mean}^{obs} is the arithmetic mean of the observed values, and Q_{mean}^{sim} is the arithmetic mean of the observed values can be computed to evaluate the skill of the model at producing correct relative peaks of simulated streamflow. Correlation coefficient varies from –1 to 1, providing information about variables that are negatively correlated ($r = -1$), uncorrelated ($r = 0$), and positively correlated ($r = 1$). It is a useful diagnostic for evaluating a hydrologic model that is being used for flood forecasting. The normalized bias is computed as

$$NB = \frac{\sum_{i=1}^{n} Q_i^{obs} - Q_i^{sim}}{\sum_{i=1}^{n} Q_i^{obs}} \tag{7.25}$$

where Q_i^{obs} is the ith observed value and Q_i^{sim} is the ith simulated value. The NB is useful for quantifying how the hydrologic model performs in terms of the overall volume of water. Bias is useful in water management scenarios where the goal is to match the volume of water present over long time periods such as seasons or water years.

The statistics shown above are suitable for evaluating **continuous variables** such as Q_i^{sim}. Another set of statistics that relies on evaluating **dichotomous events** (i.e., yes/no) is also useful in a hydrologic evaluation context. The contingency table reduces observations and forecasts down to a binary set before computing statistics. Table 7.1 shows a contingency table setup for

TABLE 7.1

Contingency Table Used for Evaluating Dichotomous (Yes/No) Events

	Observed Yes	Observed No
Forecast Yes	Hits	False alarms
Forecast No	Misses	Correct nulls

all possible combinations of observed and forecast values. Note that these statistics can be computed for events that are conditioned on several thresholds. The event can thus be defined based on Q_i^{obs} exceeding one or more thresholds.

From this contingency table the probability of detection (POD) is

$$POD = \frac{hits}{hits + misses} \tag{7.26}$$

where the POD can range from 0 to 1 with a perfect score of 1 being where every observation is detected. The false alarm rate (FAR) is

$$FAR = \frac{false\ alarms}{hits + false\ alarms} \tag{7.27}$$

where FAR can range from 0 to 1 with a perfect score of 0 representing no forecasts that had no associated observations. A simple summary statistic that combines information from hits, misses, and false alarms is the critical success index (CSI) defined as

$$CSI = \frac{hits}{hits + misses + false\ alarms} = \frac{1}{\dfrac{1}{1 - FAR} + \dfrac{1}{POD} - 1} \tag{7.28}$$

where CSI ranges from 0 to 1 with a perfect score value of 1 representing no misses or false alarms.

No single statistic adequately summarizes the global performance of a hydrologic simulation. It is better practice to use several statistics to evaluate as many variables as possible under varying environmental conditions (dry and wet) to thoroughly evaluate a hydrologic model's performance. Furthermore, the statistic used for evaluation should reflect the anticipated modeling objectives of the modeler. For instance, if the model is being designed and subsequently evaluated to simulate low flow conditions for water quality modeling, then many of the described statistics including NSE would not be an appropriate measure of the model's accuracy.

7.2 Hydrological Evaluation of Radar QPE

A great deal can be learned about radar quantitative precipitation estimation (QPE) through detailed evaluations and subsequent algorithmic improvements using independent rain gauge datasets. In some remote areas, these precipitation datasets may be sparse or unavailable altogether. Or the algorithms may have incorporated the gauges as a bias adjustment into the radar-based algorithms; thus they aren't independent. Furthermore, as discussed in detail in Chapter 2, gauges have their own set of errors and collect observations at scales several orders of magnitude below that of a radar pixel. Some studies have found benefit in evaluating the performance of radar QPE algorithms from the hydrologic modeling perspective (Gourley and Vieux 2005). This is an application-oriented approach, which judges the improvements in the radar QPE algorithms based on their end use as an input to a hydrologic model. Basins (and distributed hydrologic models) incorporate the spatial variability of rainfall and subsequent streamflow integrated over space and time, and thus include the combined effect of multiple radar pixels. The benefit of a hydrologic evaluation must be balanced by the uncertainty in estimating how much of the rainfall-driven water does not translate into streamflow measured by the stream gauge at the basin outlet. This includes evapotranspiration, infiltration, storage by the soils, aquifer storage, reservoir storage, water diversions for irrigation, snowmelt, etc.

7.2.1 Case Study in Ft. Cobb Basin, Oklahoma

Gourley et al. (2010) conducted a radar QPE versus rain gauge evaluation using the U.S. Department of Agriculture (USDA) Agricultural Research Service's Micronet located in the Ft. Cobb watershed in west central Oklahoma (Figure 7.9).

The primary emphasis of the study was to conduct a hydrological evaluation of dual-polarization rainfall algorithms to supplement radar QPE-gauge comparisons. This evaluation sheds light on the propagation of errors from QPE to streamflow simulations. For instance, it would be informative to hydrologic modelers and QPE developers alike if QPE random errors cancel each other out during the space-time integration of rainfall transformed into streamflow and thus have no impact on hydrologic simulation. Heavy rainfall events were collected over several years, totaling 1299 radar-gauge pairs, and several algorithms for generating quantitative precipitation estimates from the dual-polarized radar KOUN located in Norman, Oklahoma, were evaluated. The algorithms had varying levels of complexity and contain information from radar reflectivity Z_h, differential reflectivity Z_{DR}, as well as specific differential phase K_{DP}. Figure 7.10 shows scatterplots (in log scale) for all hours with measurable precipitation during the heavy rainfall events over the basin. The primary finding was the RMSE decreases and the correlation coefficient increases with increasing algorithm complexity illustrating the advantage of more advanced QPE algorithms available when utilizing dual-polarization radars.

Figure 7.11 shows the behavior of the normalized bias (*NB*) for each event comprising the dataset. The *R(Z)* algorithm reveals the greatest storm-to-storm variability in terms of *NB*. The *R(Z)* algorithm underestimates rain gauge amounts by ~40% during the tropical storm Erin case

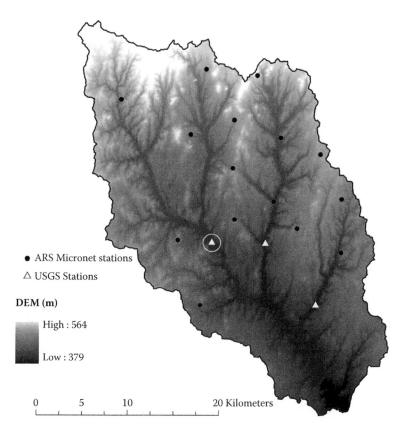

FIGURE 7.9
The Ft. Cobb basin in west central Oklahoma. The drainage area associated to the catchment measured by the stream gauge that is circled is 342 km².

on 8/18/07 and then switches to overestimation by ~80% during a spring thunderstorm event on 3/31/08. The difference between these two events lies in the variability of the drop size distribution (DSD). Tropical storm Erin had anomalous DSDs that deviated from the assumed DSD in the $R(Z)$ relation. Tropical DSDs are characteristic of a shift to a larger proportion of small-diameter drops; this causes underestimation in a convective or stratiform $R(Z)$ algorithm (Petersen et al. 1999). On the other hand, the 3/31/08 case was associated with hail reports. This meant the observed DSD was again anomalous, but shifted toward much larger diameter particle sizes. The NB with a single-parameter radar QPE algorithm thus strongly depends on the behavior of the DSDs and how these differ from the underlying assumed DSD used in the $R(Z)$ equation.

All the polarimetric QPE algorithms reveal less event-to-event variability than $R(Z)$ in terms of the NB. This indicates that the polarimetric variables Z_{DR} and K_{DP} are capable of responding to the DSDs and adjusting the

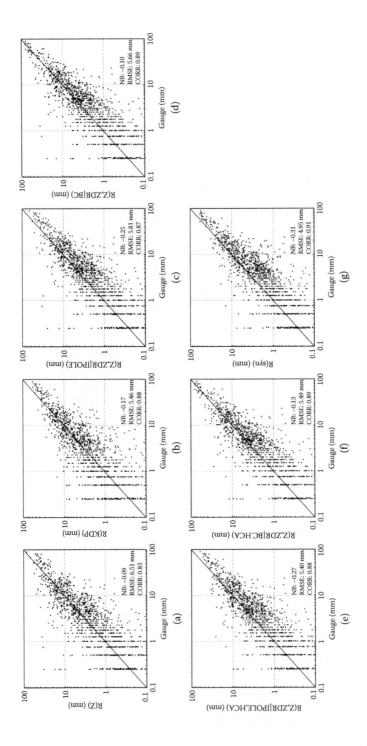

FIGURE 7.10

Scatterplots of hourly rainfall from KOUN (a) *R(Z)*, (b) *R(KDP)*, (c) *R(Z,ZDR|POLE)*, (d) *R(Z,ZDR|BC)*, (e) *R(Z,ZDR|POLE,HCA)*, (f) *R(Z,ZDR|POLE,HCA)*, and (g) *R(syn)* compared with collocated ARS Micronet rain gauge accumulations. (Figure adapted from Gourley et al. 2010.)

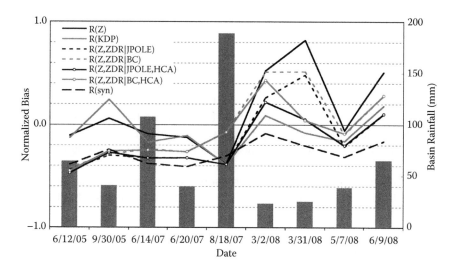

FIGURE 7.11

Normalized bias (*NB*) of the evaluated rainfall estimates for each event. Notice how the R(syn) algorithm has an overall *NB* of –0.31 (–31%), but it has the least variability from storm to storm. (Figure adapted from Gourley et al. 2010.)

rainfall estimates accordingly. The most complex "synthetic" algorithm *R(syn)* relies on thresholds to adjust the estimators so as to minimize the measurement errors from the polarimetric variables (Ryzhkov et al. 2005). For instance, at S band, K_{DP} is known to have little sensitivity in light rain. However, it is insensitive to hailstones that are mixed in with heavy rain; this mixed hydrometeor situation creates inflated *Z* values. Therefore, *R(syn)* relies on *R(K_{DP})* for heavy rainfall rates, but not at light rainfall rates. The overall *NB* with *R(syn)* was –31%, but this bias was virtually independent of storm types and rainfall intensities, indicating its ability to adapt to different DSDs.

7.2.2 Evaluation with a Hydrologic Model Calibrated to a Reference QPE

In the Ft. Cobb basin, a very dense network of high-quality rain gauges (see Figure 7.9) undergoes regular maintenance and manual quality control. These features of the rain gauge network permit their use as a reference QPE product to not only directly evaluate the QPE algorithms as shown in Figures 7.10 and 7.11, but also to calibrate a distributed hydrologic model to be used for the hydrologic evaluation component. Point values from the dense gauge network are spatially interpolated to yield a gridded QPE that is assumed to be very close to the true rainfall for the basin in terms of magnitude and spatiotemporal distribution. Furthermore, this gridded QPE field is independent from the other rainfall estimators that use the polarimetric radar variables.

The reference QPE field *R(gag)* is input to a distributed hydrologic model set up over the Ft. Cobb basin. Then, a method called Assessment of Rainfall Inputs using DREAM (ARID) is developed to first calibrate model parameters using the DREAM automatic parameter estimation method. At this point, the outcome is a distributed hydrologic model that has been calibrated to the reference rainfall, which is closest to the true rainfall. The next step is to evaluate the different QPE algorithms by inputting them to the calibrated hydrologic model and evaluating the resulting streamflow simulations. The evaluation is not absolute in that the hydrologic simulations using the radar QPEs are not expected to yield improved skill over the *R(gag)* input that was used for model calibration. It is thus useful to employ the following relative statistical measures that evaluate the simulations in relation to observed streamflow as well as the *R(gag)*-forced simulation:

$$
GRB = \left[\frac{\sum_{i=0}^{N}\left(Q_i^R - Q_i^{obs}\right)}{\sum_{i=0}^{N} Q_i^{obs}} - \frac{\sum_{i=0}^{N}\left(Q_i^{R(gag)} - Q_i^{obs}\right)}{\sum_{i=0}^{N} Q_i^{obs}} \right] \times 100 \tag{7.29}
$$

where *GRB* is the gauge-relative bias, Q^R is streamflow for the rainfall algorithm being evaluated, Q^{obs} is the observed streamflow, and $Q^{R(gag)}$ is streamflow from the R(gag)-forced simulation. The GRB thus computes the bias (in %) in relation to the bias that was present with the calibrated simulation from R(gag). A *GRB* of 0% indicates the simulation bias was the same as that achieved with *R(gag)* inputs. Similarly, the gauge-relative efficiency (GRE) is defined as follows:

$$
GRE = 1 - \frac{\sum_{i=0}^{N}\left(Q_i^R - Q_i^{obs}\right)^2}{\sum_{i=0}^{N}\left(Q_i^{R(gag)} - Q_i^{obs}\right)^2} \tag{7.30}
$$

This formulation is quite similar to the NSE in Equation (7.23), but it quantifies the skill in relation to the reference simulation from *R(gag)* forcing instead of the average of the observed streamflow. A *GRE* score of 0 indicates the *R* rainfall input resulted in the same skill that was obtained using *R(gag)* forcing, while a score of 1 indicates the simulation skill exceeded that produced during model calibration and agreed perfectly with observations. *GRE* scores worsen as they become more negative up to −∞.

The study goes on to use the ARID framework to evaluate the hydrologic performance of the various polarimetric radar algorithms. Figure 7.12 illustrates the algorithm hydrologic skill in terms of *GRE* and *GRB*. The color-filled circles

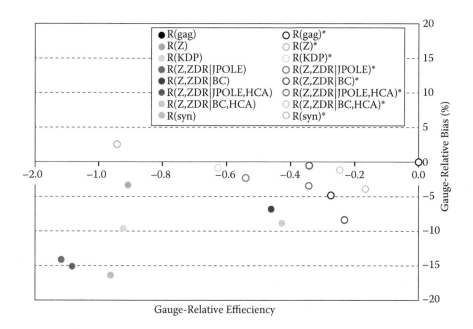

FIGURE 7.12
Hydrologic skill of polarimetric rainfall algorithms on the Ft. Cobb basin based on the *GRE*
and *GRB* metrics in Equations (7.29) and (7.30). (Figure adapted from Gourley et al. 2010.)

indicate statistical performance of the algorithms as is, whereas the circles
without color filling show the skill after the event-combined bias was removed.
This bias removal had almost no impact on the *R(Z)* algorithm's hydrologic skill
in terms of GRE. This means that *R(Z)* is unbiased when considering all events
combined, but has a great deal of variability from event-to-event. The *R(syn)*
algorithm had a lower *GRE* score than *R(Z)* inputs. In this case, the simple bias
removal improved the *GRE* so that it was only slightly less skillful than the
R(gag) inputs used during calibration. Moreover, following bias adjustment, the
GRE skill of the inputs generally followed the complexity of the polarimetric
algorithms themselves in terms of conditional use of the variables. This means
that polarimetric radar is capable of providing inputs for hydrologic modeling
that are more precise, but can be prone to systematic errors, or bias. This study
points to the great potential of using polarimetric radar, but also cautions on
the need to correct variables that may be more sensitive to bias.

7.2.3 Evaluation with Monte Carlo Simulations from a Hydrologic Model

It is not always feasible to study the hydrology of a basin that has reference
rainfall and streamflow measurements that can be considered the ground
truth. Another approach to hydrologic evaluation of precipitation inputs is to
consider the largest contributions to uncertainty within the hydrologic model.

Typically, this is the parametric uncertainty. One method for accounting for uncertainty in hydrologic model parameters is to perform many simulations in a Monte Carlo fashion and explore a portion or all of the available parameter space. From this number of simulations the resulting skill of all of the members can be evaluated and put into an appropriate statistical context. This method eliminates the need for conventional hydrologic model calibration but provides only ranges for the potential hydrologic skill of a given precipitation input. If the precipitation inputs are but slight modifications of one another, it is possible that the uncertainty in the hydrologic skill will be driven by the parametric uncertainty, making it difficult to distinguish the skill of one precipitation input from the skill of another.

Gourley and Vieux (2005) demonstrated an example of Monte Carlo simulations for evaluating the skill of several radar-based inputs. The parameter space of the hydrologic model used is uniformly sampled and probability density functions (pdfs) of peak streamflow, time at which the peak occurred, and total water volume were computed. The pdfs are nonparametric and are derived using the Gaussian kernel density estimation technique introduced in Equation (2.1). The ensemble skill is assessed by using the *ranked probability score* (*RPS*; Wilks 1995):

$$RPS = \sum_{m=1}^{J} \left[\left(\sum_{i=1}^{m} y_i \right) - \left(\sum_{i=1}^{m} o_i \right) \right]^2 \qquad (7.31)$$

where y_i is the cumulative probability assigned to the ith category, o_i is the cumulative probability of the observation in the ith category, and J is the number of categories. This score essentially compares the entire pdf of the simulated variables to a single observation of time, peak, or volume. Simulations that are far removed from the observation are penalized more heavily than those falling into nearby categories. When comparing statistical scores, it is often necessary to establish the statistical significance of their differences. In the above study, a resampling technique was employed to establish the confidence intervals. This was accomplished by pooling together the different samples of hydrograph derivatives (e.g., peak streamflow). Then, the values are randomly chosen to create two different samples. RPS scores are computed from each sample 1000 times and subsequently used to create cumulative distributions of the *RPS* differences. The cumulative distribution is then used to determine the probability of obtaining the original *RPS* difference and thus serves as the basis for computing the statistical significance of *RPS* differences obtained from simulations forced by different rainfall algorithms.

For this method to be successful, the sensitive parameters of the hydrologic models must be identified and their distributions need to be approximated. The sampling of the parameter space is performed uniformly. If only a portion of the parameter space is sampled, and this leads to biased simulations, then the technique will incorrectly identify a given precipitation input as the best in terms of hydrologic skill. Nonetheless, this method is useful for quantifying and comparing the hydrologic skill of different rainfall inputs when a true rainfall reference is not known. Furthermore, the method samples the parameter space rather than finding an optimum location that results in the best hydrologic skill. This means the method essentially avoids parameter estimation and can thus be employed on a basin that does not have long records of continuous precipitation. Generally, a hydrologic model requires years of continuous precipitation records in order to calibrate parameters; this is a condition that generally limits the application of hydrologic evaluation methodologies that rely on model calibration. They key results from the study conducted on the Blue River basin in Oklahoma are that sparse rain gauge-based inputs do not perform as well as radar-based inputs in a distributed hydrologic model. And, similar to the study conducted on Ft. Cobb, they generally found better hydrologic skill with increasing complexity in the precipitation algorithms.

7.2.4 Evaluation with a Hydrologic Model Calibrated to Individual QPEs

Another technique for evaluating the skill of individual precipitation estimates is to calibrate the hydrologic model separately for each precipitation estimate. This assumes that long time periods of the different precipitation estimates, on the order of years, are available for calibrating the hydrologic model and that the parameter estimation technique performs similarly and objectively for each input. Obtaining long, continuous precipitation estimate records can pose serious limitations to this approach. This is especially the case for remote-sensing algorithms that are continually undergoing updates due to improved estimation approaches and new technological advances in the observing platforms. The complex nature of hydrologic models means that parameter estimation methods must attempt to sample the entire parameter space or a good approximation of it. Given that the parameter space is often 10 or more dimensions, it is not possible to completely sample the entire parameter space with modern computational power. Many automatic parameter estimation methods thus identify local minima in the multidimensional parameter space. The representativeness of this local minimum can vary depending on the input. Therefore, the objectivity of the evaluation can become compromised by the capability of the optimization procedure, rather than delivering a hydrologic skill of the precipitation forcings.

Because of the difficulty in securing long records of the precipitation estimates, especially from nonoperational systems, and because of the possibility that the optimization method is not identifying the true global minimum, this technique is not often employed in research. However, if long records show that precipitation inputs have not undergone significant changes, then this technique can be quite useful for operational hydrologic forecast systems. It is possible that a given precipitation input results in better hydrologic skill because it is compensating for a deficiency in the hydrologic model structure that would ordinarily result in biased simulations.

Thus, if an operational user desires a better hydrologic forecast, then this technique will inform them of the product that accomplishes the goal of improved hydrologic forecast capability. The same conclusion may not directly apply to the absolute skill of the rainfall algorithm itself and the results will be limited to the specific basin, model, and time period employed.

Problem Sets

CONCEPTUAL/UNDERSTANDING PROBLEMS

1. Describe the key difference, in terms of parameters, from a conceptual hydrologic model and physically based hydrologic model.
2. Describe the differences and similarities between lumped, semidistributed, and distributed hydrologic models.
3. Why is it difficult to accurately evaluate the performance of precipitation inputs using hydrologic models?
4. How does evaluating a QPE using a hydrologic metric better account for the spatiotemporal scales of the QPE?
5. What are three hydrologic variables that can be evaluated?
6. What is meant by error propagation in terms of radar QPE and hydrologic models?

QUANTITATIVE EXERCISES

1. Given a contingency table with 120 hits, 38 misses, 45 false alarms, and 80 nulls, calculate the POD, FAR, and CSI.
2. Run the lumped CREST model and plot a hydrograph using the time series data provided from the Ft. Cobb watershed along with the NSCE, bias, and root-mean-square error statistics.
3. Run the lumped CREST model with precipitation input biased 10%, 25%, 50%, 75%, 100% and plot a figure showing the error propagation. Discuss how hydrographs and skill scores differ for the various QPEs. Which one has the the smallest error propagation ratio?

References

Clark, M., D. Rupp, R. Woods, X. Zheng, R. Ibbitt, A. Slater, J. Schmidt, and M. Uddstrom. 2008. Hydrological data assimilation with the ensemble Kalman filter: Use of streamflow observations to update states in a distributed hydrological model. *Advances in Water Resources* 31: 1309–1324 10.1016/j.advwatres.2008.06.005.

Duan, Q. Y., V. K. Gupta, and S. Sorooshian. 1993. Shuffled complex evolution approach for effective and efficient global minimization. *Journal of Optimization Theory and Applications* 76 (3): 501–521.

Evensen, G. 1994. Sequential data assimilation with a nonlinear quasi-geostrophic model using Monte Carlo methods to forecast error statistics. *Journal of Geophysics Research* 99: 10143–10162 10.1029/94jc00572.

Gourley, J. J., and B. E. Vieux. 2005. A method for evaluating the accuracy of quantitative precipitation estimates from a hydrologic modeling perspective. *Journal of Hydrometeorology* 6 (2): 115–133.

Gourley, J. J., S. E. Giangrande, Y. Hong, Z. L. Flamig, T. J. Schuur, and J. A. Vrugt. 2010. Impacts of polarimetric radar observations on hydrologic simulation. *Journal of Hydrometeorology* 11: 781–796.

Hamill, T. M. 2006. Ensemble-based atmospheric data assimilation. In *Predictability of Weather and Climate*, edited by T. Palmer and R. Hagedorn. Cambridge: Cambridge University Press.

Jazwinski, A. 1970. *Stochastic Processes and Filtering*. New York: Dover Publications.

Kalman, R. E. 1960. A new approach to linear filtering and prediction problems. *Journal of Basic Engineering* 82: 35–45.

Koren, V. I., M. Smith, D. Wang, and Z. Zhang. 2000. Use of soil property data in the derivation of conceptual rainfall-runoff model parameters. *Proceedings of the 15th Conference on Hydrology*, 103–106, Am. Meteorol. Soc., Long Beach, CA.

Moore, R. J. 1985. The probability-distributed principle and runoff production at point and basin scales. *Hydrological Sciences* 30: 273–297.

Liang, X., D. P. Lettenmaier, E. F. Wood. 1996. One-dimensional statistical dynamic representation of subgrid spatial variability of precipitation in the two-layer variable infiltration capacity model. *Journal of Geophysics Research* 101(D16): 21403–21422.

Nash, J., and J. Sutcliffe. 1970. River flow forecasting through conceptual models. Part I: A discussion of principles. *Journal of Hydrology* 10: 282–290.

Petersen, W. A., et al. 1999. Mesoscale and radar observations of the Fort Collins flash flood of 28 July 1997. *Bulletin of the American Meteorological Society* 80: 191–216.

Ryzhkov, A. V., S. E. Giangrande, and T. J. Schuur. 2005. Rainfall estimation with a polarimetric prototype of WSR-88D. *Journal of Applied Meteorology* 44: 502–515.

Schumann, G. P., J. C. Neal, D. C. Mason, P. D. Bates. 2011. The accuracy of sequential aerial photography and SAR data for observing urban flood dynamics: A case study of the UK summer 2007 floods. *Remote Sensing of Environment* 115 (10): 2536–2546. http://dx.doi.org/10.1016/j.rse.2011.04.039.

Vrugt, J. A., C. J. F. ter Braak, C. G. H. Diks, B. A. Robinson, and J. M. Hyman. 2009. Accelerating Markov chain Monte Carlo simulation by differential evolution with self-adaptive randomized subspace sampling. *International Journal of Nonlinear Sciences and Numerical Simulation* 10: 273–290.

Wang, J., Y. Hong, L. Li, J. J. Gourley, S. I. Khan, K. K. Yilmaz, R. F. Adler, F. S. Policelli, S. Habib, D. Irwin, A. S. Limaye, T. Korme, and L. Okello. 2011. The coupled routing and excess storage (CREST) distributed hydrological model. *Hydrological Sciences Journal* 56: 84–98. doi: 10.1080/02626667.2010.543087.

Whitaker, J. S., and T. M. Hamill. 2002. Ensemble data assimilation without perturbed observations. *Monthly Weather Review* 130: 1913–1925.

Zhao, R., Y. Zhang, L. Fang, X. Liu, and Q. Zhang. 1980. The Xinanjiang model. Hydrological forecasting, in *Proceedings of Oxford Symposium*, 129 (IAHS Publication, Wallingford, UK), 351–356.

Zhao, R. J. 1992. The Xianjiang model applied in China. *Journal of Hydrology* 135 (3): 371–381.

8

Flash Flood Forecasting

Among all storm-related hazards in the United States and beyond, flash flooding has been the deadliest in recent years (Ashley and Ashley 2008). Unlike tornadoes, large hail, and heavy snow, flash flooding is not a purely atmospheric phenomenon. Meteorologists and weather forecasters tasked with predicting these events must possess not only meteorological knowledge but also, in many cases, hydrologic knowledge as well as an understanding of the land-atmosphere interactions involved. Moreover, the magnitude and types of impacts caused by flash flooding are often dictated by behaviors, and thus social science also plays a role. Because flash floods occur, by definition, over small spatial and temporal scales, high-resolution observations are needed to produce useful forecasts.

When considering the meteorological component, high rainfall rates are the central components in producing flash flooding. Doswell et al. (1996) discuss the other atmospheric ingredients of flash flood forecasting in depth, but factors including high precipitation efficiencies, deep convective storm complexes, slow or training storm motions, and continued entrainment of moist air into storms are all significant. Over the mesoscale time and space scales, forecasters often consider the magnitude and persistence of vertical motion, precipitable water values, and high lapse rates. However, the main thrust of flash flood forecasting has and continues to be on the storm scale, typically after rainfall has already begun.

Weather radar is the most important tool for monitoring heavy rainfall that precedes flash floods. Using raw rainfall rate estimates from weather radar presents challenges to the flash flood forecaster, however. Errors in radar rainfall rates, discussed in detail in Chapter 2, can mislead a forecaster. Many flash flooding events have been associated with warm rain events (Petersen et al. 1999). If the rainfall estimation algorithm does not incorporate information about drop size distributions from polarimetric radar or recognize the signatures of a tropical environment, then severe underestimation results. The Multi-Radar/Multi-Sensor System (MRMS), discussed in Chapter 4, is well suited to accurately estimate rainfall rates in flash-flooding situations. Individual radar sites produce new volume scans roughly every 5 min, but these update times are not in sync across the network. In other words, one radar site may update at 12:05 UTC, but its neighbor may have new data available at 12:06 UTC. MRMS leverages these time differences to produce updated precipitation estimation products at a resolution of 1 km every 2 min. With this, the high resolution is maintained even after the data have been mosaicked, thus enabling rainfall rates estimated at the flash flood scale.

Flash flood forecasting requires more than just atmospheric information. For example, 25 mm of rainfall falling over a field of corn in 30 min will have a much different impact than the same amount of rainfall falling over downtown New York City in 30 min. This simple example illustrates the need for the forecaster to possess information about land surface conditions. In the New York City example, the key difference between the cornfield and the city is the amount of impervious area. When rainfall lands on plants and soils, a percentage of it infiltrates into the Earth's surface. Any water that infiltrates will not immediately run off across the land surface; this reduces the impact of immediate flash flooding. When rainfall lands on the roofs of buildings, or on any concrete or asphalt surface, none of the water can infiltrate and it immediately becomes runoff, increasing the likelihood of a flash flood. The amount of infiltration depends upon antecedent soil moisture conditions, the type of land cover, and other factors. These factors are traditionally used in hydrologic models as discussed in Chapter 7, which are designed to provide information about where water goes after it reaches the surface of the Earth.

Precipitation estimates can be used directly in hydrologic models to provide guidance about flash flooding potential. In the future, forecasters may directly use **quantitative precipitation forecasts (QPFs)** from numerical weather prediction models in hydrologic models to estimate flash flooding impacts hours or days before an event. Currently, QPE is used in hydrologic models so that the only true "forecast" component comes from the land surface component, not from the atmospheric component. For the past several decades, the U.S. National Weather Service (NWS) has produced a suite of products called **flash flood guidance (FFG)** that serves to combine hydrological and atmospheric information into one metric.

8.1 Flash Flood Guidance

In the United States, river (fluvial) flood forecasting is the responsibility of river forecast centers (RFCs) and flash flood monitoring and prediction is the responsibility of local weather forecast offices (WFOs) (Figure 8.1).

This division of responsibility is based on the amount of time it takes for rainfall over a basin to cause flooding impacts in that same basin. Floods that take more than six hours to develop are considered the responsibility of RFCs, and those that take less than six hours are defined as flash floods. Typically local WFOs have at most only one hydrologist on staff. Because forecasting flash flooding requires hydrological knowledge, this staff hydrologist (or service hydrologist) has an important responsibility to provide resources

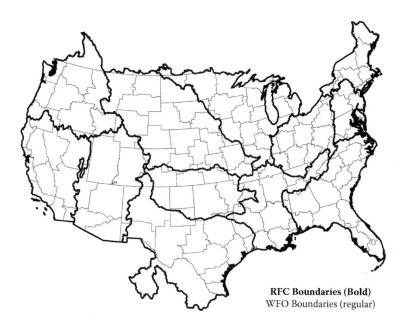

FIGURE 8.1
Map of River Forecast Center borders (bold) and Weather Forecast Office borders (regular).

to the on-duty WFO meteorologists. The regional RFCs employ multiple hydrologists who can also serve as knowledge conduits to the WFOs. As part of their normal duties, the RFCs run different hydrologic models at regular intervals for the purposes of monitoring stream stage and flow on large river networks. These same hydrologic models, with some modifications, are used to produce flash flood guidance. FFG values are produced at the RFC level and then provided to local WFOs; the values are the primary source of model information employed in the flash flood forecasting process.

Flash flood guidance is defined as the amount of rainfall required in a given period of time (1, 3, 6, 12, or 24 hrs) to induce bankfull conditions on small natural stream networks. The FFG product is plotted on a polar stereographic grid with a nominal resolution of 4 km on a side. Although it is possible for each grid cell to have an independent and separate FFG value, the spatial variability of the product depends heavily on the topography and soil type of the area in question, as well as the method used to produce FFG values. In operations, rainfall estimates from WSR-88D radars are compared in near real time to a grid of FFG values. A program called FFMP (Flash Flood Monitoring and Prediction) serves as the framework in which these comparisons are made. Forecasters can see the ratio of QPE to FFG, the difference between QPE and FFG, or the precipitation amounts expressed as a percentage of FFG. When QPE begins to approach the FFG value for a certain grid cell, forecasters may choose to issue a flash flood warning.

In some parts of the country, QPE exceeding FFG is used as a strict threshold for issuing a warning. In other areas, forecasters may wait for QPE to reach 125% or 150% of FFG before doing so. Although FFG is not intended for use in urbanized areas, some regions will fill in FFG values over urbanized grid cells. Typically this process is accomplished using some common rule of thumb, like one inch (25 mm) of rain per hour. This adjusted urban FFG is a simple way of adding urbanization information to the hydrologic model output. Clark et al. (2014) assessed the skill of various FFG products across the United States and determined that the product has its highest utility when warnings are issued at 125%, 150%, or 200% of guidance. Overall, raw FFG has much lower skill than flash flood warnings, which suggests that local knowledge or other information added to the warning process by forecasters is significantly improving forecast outcomes.

Flash flood guidance is typically produced between one and four times per day and is valid at either the synoptic (00 or 12 UTC) or subsynoptic (06 or 18 UTC) times. Each RFC transmits its FFG grid to the WFOs wholly or partially located in its area. In situations where heavy rainfall and/or flash flooding are ongoing, WFOs can request updates of FFG from their RFC in between typical issuance times. The process of updating FFG values can cause undesirable side effects like the "jump." Consider a FFG grid valid at 12 UTC; over our basin of interest, the 3 hr FFG value is 2.5 inches (63.5 mm). Between 17 and 18 UTC, the grid cells in this basin receive 2.0 inches (50.8 mm) of rainfall. At 17:59 UTC, therefore, the FFMP display would show that QPE has reached 80% of 3 hr flash flood guidance. The RFC will release an updated FFG grid at 18 UTC; the new FFG value will be around 0.5 inch (12.7 mm) (because 2.5 inches or 63.5 mm were originally needed and 2.0 inches or 50.8 mm has fallen recently). However, when the new FFG grid is loaded into FFMP, the program now indicates that rainfall is 400% of flash flood guidance! This is because the 2.0 inches (50.8 mm) of rain that have already fallen are now being compared with an updated FFG value of 0.5 inch (12.7 mm). This is one of the many caveats involved in interpreting flash flood guidance products. We refer to the phenomenon as the "jump" because a plot of QPE-to-FFG ratio between 17 and 19 UTC would go from 0% to 80% from 17 to 18 UTC and then immediately jump to 400% at 18:01 UTC.

Some RFCs issue FFG only one time a day (typically at 12 UTC), particularly during dry times of year when the FFG values tend to remain fairly constant. WFOs may also be responsible for counties across areas of two or even three RFCs. Because different generation methods are used to produce FFG at different RFCs, this can result in drastic changes in FFG values from county to county and especially from region to region. There are also grid cells along the boundaries between RFCs for which FFG values are never produced. In Figure 8.2, dark areas indicate pixels where FFG values were only sporadically available between 2006 and 2010; RFC boundaries, missing basins and pixels, and other issues are readily apparent.

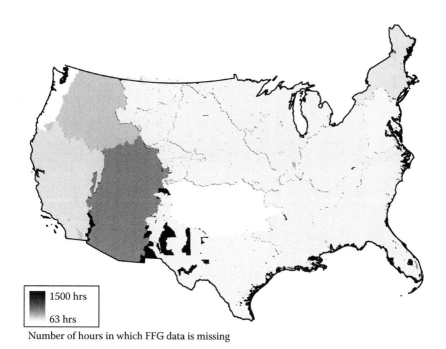

Number of hours in which FFG data is missing

FIGURE 8.2
Number of hours FFG values were available per pixel between October 2006 and August 2010. (After Clark et al. 2014.)

Flash flood guidance traditionally consists of two main components: a rainfall-runoff model operating in scenario mode and a fixed value called the **threshold runoff (or ThreshR)**. ThreshR is determined by surveys of small, natural stream channels in a particular area. The ThreshR is defined as the ratio of a basin's flood flow to its unit hydrograph peak. In other words, the ThreshR is the amount of runoff required to cause bankfull conditions at a particular location. Calculating ThreshR is easiest in locations with a stream gauge, but the small scale of flash flood basins and the large area of the United States make gauging even a small percentage of flash flood basins impractical. Therefore, over several decades, RFCs have sent out survey teams to determine ThreshR values for various streams. The ThreshR is a function of the geomorphological characteristics of a particular basin and does not change with the weather conditions. Because ThreshR is a point value used in a gridded product (FFG), various methods to grid ThreshR have been developed over the years. In some RFCs, the average of values from a few surveys is used to assign one ThreshR value to moderately sized (300–3000 km^2) river basin. In other areas of the country, ThreshR values are assigned by state or by county. Still other locations use geographic contouring to produce grids of ThreshR values spreading out from known survey locations (Carpenter et al. 1999).

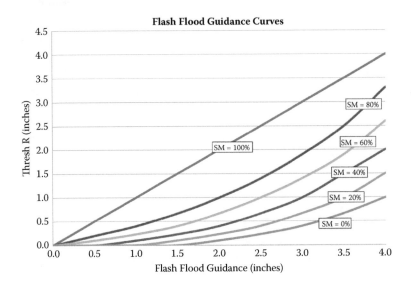

FIGURE 8.3
Curves used to produce flash flood guidance values depend heavily on soil moisture (SM). In this example, a basin with a known threshold runoff value of 1.0 in (25 mm) would have FFG values ranging between 1.0 and 4.0 in (25–102 mm), depending on the SM. As the SM value drops, the FFG value increases, because progressively more rain is needed over drier soils to produce bankfull (flooding) conditions. (After Reed and Ahnert 2012.)

The second component of FFG is the rainfall-runoff model. Figure 8.3 schematically shows the type of curves used to produce flash flood guidance values. Typically, rainfall-runoff models are forced with known rainfall amounts (QPE). The model then determines where the rainfall goes once it reaches the surface of the Earth; in other words, the model output is runoff, discharge, or streamflow. Flash flood guidance reverses this process. In FFG production, the required output is already known: the threshold runoff. The model is run through scenarios with increasing amounts of rainfall, and the amount of rainfall that led the model to produce the threshold runoff as output becomes the FFG value for that grid cell.

8.2 Flash Flood Guidance: History

Flash flooding has long been recognized as a problematic weather phenomenon. In the post–World War II years, as Americans settled new parts of the country and population densities increased in already settled areas, the number of people susceptible to flash flooding increased. By the middle 1970s, monetary damage from flash floods was six times higher than it had been immediately after the war, and three times as many people were dying from flash flooding than had been in the mid-1940s (Mogil et al. 1978). As late as 1970, the NWS did not have a national flash flood warning program, despite the existence of similar programs for severe thunderstorms and tornadoes. Several devastating flash floods at the end of the 1960s and the beginning of the 1970s spurred the formation of the flash flood warning program and with it, the first attempts at generated flash flood forecasts: "original flash flood guidance," or OFFG (Clark 2012). This OFFG product was based on anteced-ent rainfall and geomorphological characteristics of basins, just like modern flash flood guidance. In those days, national QPF forecasts regularly came from the National Meteorological Center but were of low resolution, included only synoptic-scale rainfall, and had limited utility for forecasting small-scale convective flash flood events. Warning procedures were implemented locally based on RFC expertise but could vary widely from region to region. Regular precipitation estimates were available from those areas that had weather radars, but even the use of the earliest $Z–R$ relationships required manually digitizing radar data or using the radar digitizer and processor (RADAP) technique, which was available at only a handful of radar sites in the 1970s. Towns, cities, and counties sometimes ran their own flash flood warning and alarm systems, using tornado and civil defense sirens with a special flood alert tone. Despite all these other avenues, Mogil et al. (1978) (p. 693) note that flash flood guidance was already the "critical element in local programs."

Throughout the 1970s, '80s, and early '90s, problems with OFFG appeared. Procedures to determine threshold runoff values varied from region to region, and some regions never documented how the values were produced. Sweeney (1992) explains that one RFC that covered multiple states calculated ThreshR at only four locations in their entire domain. Forecasts for large river systems were moved onto the new NWS River Forecast System in the 1980s, and in the years after that deployment, various attempts were made to move FFG generation into that framework, as well. The primary goal of these efforts was to smooth out regional differences in FFG production and to pro-vide a more consistent product to local forecasters.

The version of FFG that resulted from these efforts is today known as lumped flash flood guidance (LFFG). LFFG was the first FFG product to be produced on a grid, which allows for values to be compared with precipitation estimates on a cell-by-cell basis. Despite the gridded nature of the product, LFFG values were the same within specific river basins, usually between 300 and 5000 km^2 in size. Within these river basins, QPE-to-FFG comparisons showed significant cell-to-cell variation because high-resolution WSR-88D precipitation estimates were available in most of the country. In some regions, LFFG was an improvement over OFFG in terms of spatial resolution (Sweeney and Baumgardner 1999), but even a 300 km^2 basin is many times larger than a single pixel of QPE from weather radar. Attempts, therefore, were soon made to improve the spatial resolution of the product.

Geographic information system (GIS) software is of great utility in flash flood forecasting. As GIS technology matured, forecasters at the NWS office in Pittsburgh, Pennsylvania, recognized that GIS could be used to outline truly flash flood scale basins. They called the project AMBER, for Areal Mean Basin Estimated Rainfall. High-resolution precipitation estimates were averaged over small basins to produce better estimates of basin rainfall than those available using county-based or lumped-basin FFG. Eventually the National Basin Delineation Project (NBDP) extended this methodology across the United States (Arthur et al. 2005). These basin sizes are as small as 5 km² (Davis 2007) and rarely exceed more than 20 km² (RFC Development Management Team 2003). After the implementation of these basins was completed, rainfall estimates and basin sizes were in line with one another, but lumped flash flood guidance values were still constant over large areas.

In the years after the RFC Development Management Team issued its 2003 report, various RFCs experimented with their own versions of flash flood guidance. In the western United States, the Colorado Basin River Forecast Center developed a product called FFPI. Similar products eventually spread to the Northwest RFC and the California Nevada RFC. In 2005 and 2006, the Arkansas Red-Basin RFC developed what has become known as gridded flash flood guidance, or GFFG. As of 2010, this methodology is known to be in use in much of the south-central and southeastern United States. Finally, the Middle Atlantic RFC uses a different hydrologic model and high-resolution antecedent precipitation to produce a form of FFG called distributed flash flood guidance, or DFFG (Clark et al. 2014).

8.3 Lumped Flash Flood Guidance

Lumped flash flood guidance was developed in the early 1990s when the NWS first decided to move FFG production into the same system used for large-scale river stage forecasts. Soil moisture (and in some RFCs, snowpack) allows a hydrologic model (usually the Sacramento Soil Moisture Accounting [SAC-SMA] model, though others are available) to determine how much of a given amount of precipitation will saturate the soils and how much will run off and be available for streamflow. SAC-SMA is a lumped parameter model, so the soil moisture conditions and any model parameters are identical across a particular model basin. These model basins are more than 300 km² in size, which corresponds to time scales in excess of the six hours typically used to characterize flash floods (Clark 2012). As described earlier in the chapter, SAC-SMA is run in reverse since the threshold runoff (a hydrological model's typical output) is already known. In this case the rainfall is the unknown, and the amount of rainfall that causes the model to exceed the threshold runoff becomes the LFFG value for that entire lumped basin.

The same soil moisture used to produce river stage forecasts is used in generating LFFG (Sweeny and Baumgardner 1999). The ThreshR values used in LFFG are the ratio of flood flow over a basin to the peak of the unit hydrograph over the basin. Usually LFFG relies on the assumption that bankfull conditions are equal to a flood with a return period of two years. Survey teams calculate ThreshR at various points, especially those outfitted with stream gauges. Then those point values are contoured to create a field of ThreshR values. There are many other ways to calculate ThreshR, as described in Carpenter et al. (1999). Today, the most common method of producing ThreshR is to complete several surveys using basins outlined by digital elevation models (DEMs), contour the field of values between the survey points, and then plot the new values on the ThreshR grid (Reed et al. 2002).

The LFFG method allows for the use of models other than SAC-SMA. Differences in LFFG values across RFC domain boundaries can be partially explained if different hydrologic models are being used. Soil moisture data are updated at the RFCs every six hours (assuming that precipitation estimates are being produced in a timely manner) and this allows for LFFG to also be updated every six hours. Rapid changes in soil moisture that occur over a couple of hours or in less than an hour's time are not easily reflected in LFFG, but changes that occur over six hours or longer do appear in the product. The lumped character of the method eliminates the ability of the forecaster to easily see subbasin differences in soil moisture and antecedent precipitation. Over a 1000 km^2 basin, it is possible—if not likely—that drier parts of the basin will respond differently to new precipitation than those already near saturation. Unfortunately, LFFG largely prevents the easy examination of these differences.

Over northern and western RFCs, frozen soils, snowpack, and snowmelt is a critical part of flash flood forecasting. The Snow-17 model is used to assist in accounting for this additional water content above the soil when LFFG values are produced. It should be noted that in these snow-prone areas, LFFG values tend to exist across a much wider range than in areas that do not require monitoring of snowpack water amounts (Sweeney and Baumgardner 1999).

8.4 Flash Flood Potential Index

Flash Flood Potential Index (FFPI) is designed for areas where it is believed that soil moisture is an unimportant component of flash flood forecasting relative to geomorphological basin characteristics. The Western Region Flash Flood Analysis project was launched to create a new type of FFG for these areas (RFC Development Management Team 2003). Several layers of gridded basin parameters are resampled to identical resolutions and compared with one another in an effort to assign *relative* susceptibilities of basins to flash flooding impacts.

Smith (2003) obtained several static layers (land use, forest cover, vegetation type, slope, and soil type) and assigned each grid cell in each layer a value from 1 to 10, based on the anticipated magnitude of a flash flood response. Then he obtained dynamic layers (vegetation state, snow cover, fire, and modeled precipitation) and conducted the same process. All the values for all the layers are averaged together to yield the final relative flash flood potential for each grid cell. All layers are weighted equally except for slope, which receives a slightly higher weight. Because the values represent relative flood potentials, they are unitless. The final gridded numbers are then averaged over FFMP basins so that each basin has a susceptibility value. In some RFCs, this susceptibility value eventually replaced lumped flash flood guidance. In the Colorado Basin RFC, a 1 in/hr (25 mm/hr) rainfall rate is adjusted up or down based on the basin susceptibility value to produce a final FFG value (Clark 2012). Clark et al. (2014) determined that FFPI had little utility in flash flood forecasting, though this may be due to difficulties in obtaining accurate flash flood observations and QPE in the western United States rather than to inherent flaws in the FFPI method.

8.5 Gridded Flash Flood Guidance

Schmidt et al. (2007) describes a method of producing high-resolution flash flood guidance in an attempt to bridge the gap between extremely large lumped hydrologic model basins and small FFMP basins. This high-resolution flash flood guidance is known as gridded flash flood guidance, or GFFG. Originally developed in 2005 and 2006 over the Arkansas Red-Basin RFC (ABRFC), the method has since spread to much of the rest of the United States. The advantages of GFFG are its high resolution and its use of readily available GIS datasets. However, GFFG has some disadvantages. Until 2014, the literature did not contain any national objective evaluations of the skill of lumped flash flood guidance (Clark et al. 2014). To make GFFG easier to transition into operations, its creators purposely designed the system to contain values that mimicked those of the earlier LFFG system (Schmidt et al. 2007; Gourley et al. 2012). Therefore, any flaws in LFFG were by necessity carried forward into the new GFFG system.

A distributed hydrologic model underlies GFFG; that is, a model that has parameters that can vary from grid cell to grid cell, rather than from lumped basin to lumped basin. GFFG runs on the same 4 km polar stereographic grid used in the original LFFG product. The components of GFFG include slope (derived from a digital elevation model), soil type, land cover, soil moisture, ThreshR values, and a rainfall-runoff model. The Natural Resources Conservation Service curve number (CN) method is used to determine how susceptible each grid cell is to flooding impacts. Curve numbers are found using a lookup table that requires the user to know the land use and soil type characteristics of each grid cell. The higher the CN, the more potential for generating runoff and thus a flash flood. In GFFG, the curve numbers include information about soil moisture conditions and thus antecedent precipitation. A soil moisture model that provides a saturation ratio is used to adjust these curve numbers. This model is the HL-RDHM model (NWS Hydrology Laboratory Research Distributed Hydrologic Model), and it has two parameters of interest in this case: the upper zone free water contents and upper zone tension water contents. Koren et al. (2000) estimated the maximum possible value of each of these parameters, and so the sum of the values from the model can be divided by the sum of the maximum possible values to yield a saturation ratio.

Two equations determine the upper and lower bound of the soil moisture adjusted curve number. In the wet equation,

$$CN_{ARCIII} = \frac{23 * CN}{10 + 0.13 * CN} \tag{8.1}$$

ARCIII is what the soil moisture adjusted curve number would be if the soil was completely saturated. *CN* is the curve number under conditions of 50% saturation. The dry equation

$$CN_{ARCI} = \frac{4.2 * CN}{10 - 0.058 * CN} \tag{8.2}$$

has the term *ARCI*, which is the soil moisture adjusted curve number at 0% saturation. Therefore, the normal curve number *CN* is determined from the lookup table and is based solely on land use and soil type in the grid cell. That *CN* is substituted into Equations (8.1) and (8.2) to find the possible upper and lower bounds of the soil moisture adjusted curve number. Then the saturation ratio from the HL-RDHM model determines CN_{ARCII}, which is the final soil moisture adjusted curve number. For example, if the original curve number is 70, the saturated curve number *ARCIII* becomes 84. The unsaturated curve number *ARCI* is 49. If the HL-RDHM model determines that the grid cell is 80% saturated, the final adjusted value of the curve number *ARCII* is 77. This CN is higher than the original value of 70 and thus represents a higher potential of flash flooding, which is to be expected, since the soils were quite saturated (80%) in our example. Once the final curve number is calculated, it is used in

$$S = \frac{1000}{CN} - 10 \tag{8.3}$$

where *CN* now represents the final soil moisture adjusted curve number. In Equation (8.3), *S* is the initial abstraction (Schmidt et al. 2007).

In GFFG, threshold runoff is determined by creating a three-hour design rainfall event that has a five-year return period. The runoff from this design storm becomes the flow at flood stage on small natural stream networks. The unit hydrograph peak is determined using the NRCS curve number method and then the ratio of the flow at flood stage to the unit hydrograph peak is the threshold runoff value. Schmidt et al. (2007) discuss the changes observed in ThreshR values with GFFG and note that the primary difference is a greater range of values than would have been observed with legacy LFFG threshold runoff values. Finally, abstraction S from Equation (8.3) is used to calculate the final GFFG values

$$Q = \frac{(P - 0.2S)^2}{P + 0.8S} \tag{8.4}$$

where Q is the threshold runoff and P is precipitation. The equation must be solved for P because it is the final gridded flash flood guidance value (Schmidt et al. 2007).

Although assessments by Gourley et al. (2012) over the ABRFC and Clark et al. (2014) over the United States found little skill improvement in GFFG compared with older legacy methods of producing flash flood guidance, GFFG is able to resolve small-scale details that LFFG cannot. Most importantly, GFFG is produced at spatial scales appropriate for most flash flooding events.

8.6 Comments on the Use of Flash Flood Guidance

Documentation of updates and changes to flash flood guidance is scarce. Because of this, local forecasters have at times undertaken their own modifications of and alterations to the product. One of the most common involves manually lowering the guidance values over known urbanized areas, as needed. Other modifications involve creating smaller subdivided basins from the original National Basin Delineation Project basins. Davis (2004) explains that some flash flood guidance methods make questionable assumptions. In particular, the rainfall-runoff models run in scenario mode assume that rainfall is equally distributed in space and time. Additionally, the small natural streams modeled FFG are assumed, in some methods, to be at a low-flow condition at the start of a particular FFG issuance. Of course, if the streams already have some amount of water contained within them, FFG will overestimate the amount of rainfall required to induce flash flooding.

The division of flash flood forecasting responsibilities between the regional and local forecasters has resulted in a situation in which regular rigorous evaluation of flash flood guidance is extremely difficult. Gourley et al. (2012)

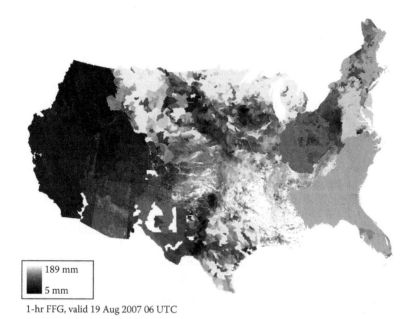

1-hr FFG, valid 19 Aug 2007 06 UTC

FIGURE 8.4
National mosaic of flash flood guidance valid at 06 UTC on August 19, 2007.

and Clark et al. (2014) have completed objective evaluation of flash flood guidance skill regionally and nationally, respectively. Schmidt et al. (2007) and Smith (2003) conducted case-by-case evaluations of their FFG modifications prior to the deployment of GFFG and FFPI, respectively. Other case evaluations of the product exist in the literature and most of these recognize the need for significant modifications to the FFG methodology. Figure 8.4 shows how the current patchwork of methods results in different FFG values from RFC to RFC. For all these reasons, caution is warranted when heavily using flash flood guidance information to issue flash flood warnings and other products.

8.7 Threshold Frequency Approach

Because flash flooding occurs at small spatial and temporal scales, attempts to directly model flows on high-resolution grids have been recently been made. One of proposed methods is known as DHM-TF (Distributed Hydrologic Modeling—Threshold Frequency) (Reed et al. 2007). The general concept of DHM-TF is similar to FFG; flood forecasts at all grid cells (including ungauged locations) are desired (Cosgrove et al. 2010). However, rather than using a rainfall-runoff model in scenario mode, DHM-TF requires a hydrologic model to be run in direct forward simulation mode. In other words, rather than iterating through rainfall scenarios and determining which depth/duration induces bankfull conditions on small streams, observed precipitation estimates (QPE) are used as input to a hydrologic model. The model then outputs discharge (or streamflow) at each grid cell. Because most of the grid cells will be ungauged, the raw streamflow outputs from the model have to be characterized in terms of their potential to cause flooding. Typically this is done by running the same hydrologic model over a lengthy period and then obtaining a distribution of past simulated flows at each grid cell. Then as the model runs forward in time, in forecast mode, the streamflow output can be compared to the historical distribution at that grid cell to determine how potentially severe the flooding impacts might be.

DHM-TF has several advantages over flash flood guidance, including increased resolution. A distributed hydrologic model can be run on almost any grid mesh, assuming precipitation, DEM, and other input data are available at reasonably similar resolutions. With the right amount of computing power, DHM-TF can provide much more frequent updates than flash flood guidance, as well. Every time new precipitation data are available (in the United States, at least every five minutes, and potentially more frequently with the new MRMS system), the model can be rerun. Of course, computational speed dictates how quickly new streamflow estimates will be available after a particular model run. This represents a significant advantage over the four times daily or once daily updates that are the current standard in the FFG system (Cosgrove et al. 2010). Unlike with GFFG, the use of a direct simulation in DHM-TF allows for water to be transferred from grid cell to grid cell using the model's built-in routing (if such capability is available) (Reed 2008). Any distributed hydrologic model can be used to complete the DHM-TF method; the prototype described by Reed et al. (2007), Cosgrove et al. (2010), and others uses the HL-RDHM model. Recall that parameters from HL-RDHM are used to derive GFFG as well. The most significant improvement over flash flood guidance, however, is that DHM-TF requires only the simulation of the *relative* importance of events. In other words, as long as the model correctly (or nearly correctly) places a particular event within historical context, the exact streamflow amount from the model is unimportant.

Gourley et al. (2014) showed that, for certain cases in the south central United States, the skill of the DHM-TF method exceeded that of legacy lumped flash flood guidance as well as the newer GFFG methodology. Their method used a different hydrologic model to generate streamflow outputs: CREST (Coupled Routing and Excess STorage). The CREST model is a joint venture of the University of Oklahoma and the National Aeronautics and Space Administration (NASA) (Wang et al. 2011). CREST also underpins a preoperational suite of DHM-TF products called FLASH (Flooded Locations and Simulated Hydrographs). The FLASH project is housed by the University of Oklahoma and the National Severe Storms Laboratory and is anticipated to be available to National Weather Service forecasters in 2016. FLASH consists of a distributed hydrologic model (CREST) running at 10 min and 1 km resolution over the conterminous United States. The model is forced with precipitation estimates from NSSL's MRMS project, discussed in detail in Chapter 4. Currently, this QPE is fed to the model, at which point the model assumes no further precipitation occurs. The model runs forward in time 6 hrs; the maximum streamflow in each grid cell over that 6 hr period is converted to a return period. The return period is determined from a Log-Pearson Type III distribution (Pearson 1895) that consists of the maximum annual flows or partial duration series at each grid cell from 2002 to 2012. Those maximum annual flows were in turn found after running the model in hindcast mode forced by Stage IV (Lin 2007) precipitation estimates. An example of a regional FLASH grid is presented in Figure 8.5.

Given flash flood guidance limitations and advanced age, distributed hydrologic models that are forced with observed or forecast rainfall (rather than rainfall scenarios) are expected to gradually become the central way of forecasting and monitoring flash flood events in the United States and beyond. The same methodology is also evolving in other countries with

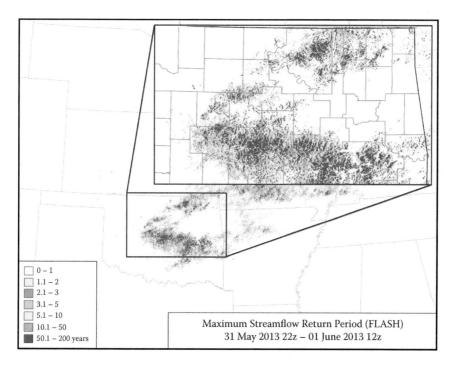

0 – 1	
1.1 – 2	
2.1 – 3	
3.1 – 5	
5.1 – 10	
10.1 – 50	Maximum Streamflow Return Period (FLASH)
50.1 – 200 years	31 May 2013 22z – 01 June 2013 12z

FIGURE 8.5
Example of a FLASH maximum return period forecast from May 31, 2013–June 1, 2013.

well-established weather radar networks. As the skill of numerical weather prediction improves, hydrologic models may eventually be forced with quantitative precipitation forecasts, which would allow for flash flood forecasts far in excess of the six hours currently available in FLASH. This would also improve the coupling of atmospheric and land-surface knowledge required to produce high-quality forecasts of flash flooding.

Problem Sets

QUALITATIVE PROBLEMS

1. Compare and contrast gridded flash flood guidance and lumped flash flood guidance.
2. What are some of the advantages and disadvantages of all types of flash flood guidance?
3. How can DHM-TF and FLASH improve upon flash flood guidance?
4. What types of meteorological factors contribute to flash flooding?

QUANTITATIVE PROBLEMS

1. You are forecasting flash floods for an area consisting mostly of dirt roads where the soil belongs to hydrologic soil group C, according to the NRCS. This yields a curve number of 87. The soil in this area is 40% saturated. Determine the soil moisture adjusted curve number.
2. Using the information from the previous problem, and for a stream with a threshold runoff of 3.0 inches (76 mm), determine the gridded flash flood guidance for this grid cell.
3. Using Figure 8.3, for a basin with a threshold runoff of 2.5 in. (63.5 mm) and with 60% saturated soils, determine the flash flood guidance value.
4. Using the National Weather Service's default Z–R relationship, determine the rainfall rate corresponding to a radar reflectivity factor of 50 dBZ (100,000 Z). How much rainfall occurs over a 3 hr period? If the 3 hr flash flood guidance is 2.5 inches (63.5 mm), what is the precipitation to FFG ratio?

References

Arthur, A., G. Cox, N. Kuhnert, D. Slayter, and K. Howard. 2005. The National Basin Delineation Project. *Bulletin of the American Meteorological Society* 86: 1443–1452.

Ashley, S., and W. Ashley. 2008. Flood fatalities in the United States. *Journal of Applied Meteorology and Climatology* 47: 806–818.

Carpenter, T., J. Sperfslage, K. Georgakakos, T. Sweeney, and D. Fread. 1999. National threshold runoff estimation utilizing GIS in support of operational flash flood warning systems. *Journal of Hydrology* 224: 21–44.

Clark III, R. 2012. *Evaluation of Flash Flood Guidance in the United States*. M.S. Thesis. University of Oklahoma.

Clark, R. A., J. J. Gourley, Z. L. Flamig, Y. Hong, and E. Clark. 2014. CONUS-wide evaluation of National Weather Service flash flood guidance products. *Weather and Forecasting* 29: 377–392. doi:10.1175/WAF-D-12-00124.1.

Cosgrove, B., S. Reed, M. Smith, F. Ding, Y. Zhang, Z. Cui, and Z. Zhang. 2010. DHM-TF: Monitoring and predicting flash floods with a distributed hydrologic model. Presentation, *Eastern Region Flash Flood Conference*, Wilkes-Barre, PA.

Davis, R. 2004. Locally modifying flash flood guidance to improve the detection capability of the Flash Flood Monitoring and Prediction program. Preprints, *18th Conference on Hydrology*, Seattle, WA, Amer. Meteor. Soc., J1.2. [Available online at http://www.ams.confex.com/ams/pdfpapers/68922.pdf.]

Davis, R. 2007. Detecting the entire spectrum of stream flooding with the flash flood monitoring and prediction (FFMP) program. Preprints, *21st Conference on Hydrology*, San Antonio, TX, Amer. Meteor. Soc., 6B.1.

Doswell III, C. A., H. Brooks, and R. Maddox. 1996. Flash flood forecasting: An ingredients-based methodology. *Weather and Forecasting* 11: 560–581.

Gourley, J. J., J. Erlingis, Y. Hong, and E. Wells. 2012. Evaluation of tools used for monitoring and forecasting flash floods in the United States. *Weather and Forecasting* 27: 158–173.

Koren, V., M. Smith, D. Wang, and Z. Zhang. 2000. Use of soil property data in the derivation of conceptual rainfall-runoff model parameters. Preprints, *15th Conference on Hydrology*, Long Beach, CA, Amer. Meteor. Soc., 103–106.

Lin, Y. 2007. Q&A about the new NCEP Stage II/Stage IV. Mesoscale Modeling Branch, Environmental Modeling Center, National Centers for Environmental Prediction. [Available online at http://www.emc.ncep.noaa.gov/mmb/ylin/pcpanl/QandA.]

Mogil, H., J. Monro, and H. Groper. 1978. NWS's flash flood warning and disaster preparedness programs. *Bulletin of the American Meteorological Society* 59: 690–699.

Pearson, K. 1895. Contributions to the mathematical theory of evolution, II: Skew variation in homogenous material. *Philosophical Transactions of the Royal Society B* 186: 343–414.

Petersen, W. A., et al. 1999. Mesoscale and radar observations of the Fort Collins flash flood of 28 July 1997. *Bulletin of the American Meteorological Society* 80: 191–216.

Reed, S. 2008: DHM-TF overview. Presentation, RFC Development Management. [Available online at http://www.nws.noaa.gov/oh/rfcdev/docs/DHM-TF-GFFG.pdf.]

Reed, S., and P. Ahnert. 2012. National Weather Service flash flood modeling and warning services. Presentation, *USACE Flood Risk Management and Silver Jackets Workshop*, Harrisburg, PA. [Available online at http://www.nfrmp.us/frmpw/docs/WORKSHOP/Allegheny/4%20-%20Thursday/1200_-_Reed_-_National_Weather_Service_FFW_Services_SJ_draft4.pdf.]

Reed, S., D. Johnson, and T. Sweeney. 2002. Application and national geographic information system database to support two-year flood and threshold runoff estimates. *Journal of Hydrologic Engineering* 7: 209–219.

Reed, S., J. Schaake, and Z. Zhang. 2007. A distributed hydrologic model and threshold frequency-based method for flash flood forecasting at ungauged locations. *Journal of Hydrology* 337: 402–420.

River Forecast Center Development Management Team. 2003. Flash flood guidance improvement team: Final report. Report to the Operations Subcommittee of the NWS Corporate Board, 47 pp. [Available online at http://www.nws.noaa.gov/oh/rfcdev/docs/ffgitreport.pdf.]

Schmidt, J., A. Anderson, and J. Paul. 2007. Spatially-variable, physically derived, flash flood guidance. Preprints, *21st Conference on Hydrology*, San Antonio, TX, Amer. Meteor. Soc., 6B.2. [Available online at http://ams.confex.com/amspdfpapers/120022.pdf.]

Smith, G. 2003. Flash flood potential: Determining the hydrologic response of FFMP basins to heavy rain by analyzing their physiographic characteristics. Report to the NWS Colorado Basin River Forecast Center, 11 pp. [Available online at http://www.cbrfc.noaa.gov/papers/ffp_wpap.pdf.]

Sweeney, T. 1992. Modernized areal flash flood guidance. NOAA Tech. Rep. NWS HYDRO 44, Hydrologic Research Laboratory, National Weather Service, NOAA, Silver Spring, MD, 21 pp. and an appendix.

Sweeney, T., and T. Baumgardner. 1999. Modernized flash flood guidance. Report to NWS Hydrology Laboratory, 11 pp. [Available online at http://www.nws.noaa.gov/oh/hrl/ffg/modflash.htm.]

Wang, J., Y. Hong, L. Li, J. J. Gourley, S. Khan, K. Yilmaz, R. Adler, F. Policelli, S. Habib, D. Irwn, A. Limaye, T. Korme, and L. Okello. 2011. The coupled routing and excess storage (CREST) distributed hydrological model. *Hydrological Sciences Journal* 56: 84–98.

Index

For Product Safety Concerns and Information please contact our EU
representative GPSR@taylorandfrancis.com Taylor & Francis Verlag GmbH,
Kaufingerstraße 24, 80331 München, Germany

Printed and bound by CPI Group (UK) Ltd, Croydon, CR0 4YY
02/05/2025
01859323-0004